THE CULTURE OF OUR DISCONTENT

No man ever looks at the world with pristine eyes. He sees it edited by a definite set of customs and institutions and ways of thinking.

Ruth Benedict, *Patterns of Culture*, 1934

Our Prozac-popping, cognitively focused, semi-alienated post-modernity is only a stage in the ongoing understanding and control of mood and character.

Andrew Solomon, *The Noonday Deamon*, 2001

Hearing voices no one else can hear isn't a good thing, even in the wizarding world.

J. K. Rowling, *Harry Potter and the Chamber of Secrets*, 1999

THE CULTURE OF OUR DISCONTENT

BEYOND THE MEDICAL MODEL OF MENTAL ILLNESS

MEREDITH F. SMALL

Joseph Henry Press
Washington, D.C.

Joseph Henry Press • 500 Fifth Street, NW • Washington, DC 20001

The Joseph Henry Press, an imprint of the National Academies Press, was created with the goal of making books on science, technology, and health more widely available to professionals and the public. Joseph Henry was one of the founders of the National Academy of Sciences and a leader in early American science.

Library of Congress Cataloging-in-Publication Data

Small, Meredith F.
  The culture of our discontent : beyond the medical model of mental illness / Meredith F. Small.
    p. cm.
  Includes bibliographical references.
  ISBN 0-309-10066-6 (hardcover)
  1. Cultural psychiatry. I. Title.
  RC455.4.E8S555 2006
  616.89—dc22

                            2006017868

Cover credit: Cover photos © Photodisc.

Printed in the United States of America

To Tim,
who keeps me sane.

OTHER TITLES BY MEREDITH F. SMALL

Female Primates: Studies by Women Primatologists

Female Choices: Sexual Behavior of Female Primates

What's Love Got to Do with It? The Evolution of Human Mating

Our Babies, Ourselves: How Biology and Culture Shape the Way We Parent

Kids: How Biology and Culture Shape the Way We Raise Our Children

CONTENTS

Introduction   1

1   The Rooted Sorrow   7
2   The Evolution of the Mind   35
3   The Minds of Monkeys   56
4   The Happy Fat   74
5   States of Mind   90
6   Running Amok in a Brain Fog   109
7   Cursed and Haunted   130
8   A Happy Ending   147

Notes   159

References   169

Acknowledgments   185

Index   187

INTRODUCTION

I'M STANDING AT THE ENTRANCE OF AN EXHIBIT HALL THE LENGTH AND breadth of a football field. Huge colorful banners hang from the ceiling, large digital talking heads announce new products on video screens, and well-dressed barkers hail passersby to enter their displays. At one stand, free neck massages are offered; at another you can get your name engraved on a free laser pen. And all the booths hawk drugs that purport to aid all types of human behaviors and moods.

This carnival atmosphere is not exactly what I expected at the annual meeting of the American Psychiatric Association (APA). I had been invited to give a guest lecture on how parenting styles in different cultures affect children and I anticipated a typical academic convention—seminars, discussions, and maybe a little free time to socialize with colleagues. I was especially looking forward to sitting in on talks about current trends in the treatment and understanding of mental illness, and having deep conversations about various therapy options and how they work. Maybe I would even hear the announcement of an exciting new discovery in psychiatry.

Instead, I found myself drawn again and again to the exhibit hall, a place of light and sound where the current culture of American psychiatry was vividly on display. Clearly, that culture is all about pharmaceuticals.

Drug ads hung from the ceiling, rose from the floor in giant displays, and covered the surfaces of booths. The ads addressed a cornucopia of negative human behaviors, and every symptom, syndrome, or condition called for pharmaceutical treatment. Apparently, stress, depression, irritation, premenstrual syndrome, fatigue, explosive anger, and disconnection from reality are all candidates for medication. According to the exhibitors, just about everyone in the country, including me, is suffering from some sort of mental illness, disease, or crippling behavioral symptom. And we could all be helped by pharmaceuticals.

As an anthropologist, I found this scene captivating. The exhibit hall was clearly a cultural statement about how mental problems are perceived and treated in Western culture. It was also a statement that mental illness has become a prominent feature of our culture. Mental illness is, in fact, one of the major health and social issues of our time; it touches every family in some way. Without much effort, I can come up with a long list of family members and friends who have been affected. Their diagnoses include clinical depression, bipolar disorder, hyperactivity disorder, addiction, psychotic break, situational depression, and anxiety, among others. I know about their diagnoses because people are open about discussing their psychological ills, and we now have a common public language to label our woes. It also seems like the number of unhappy, troubled individuals is growing. If mental illness was a communicable disease, we would probably now call it an epidemic.

After leaving the exhibit hall, I began to wonder, as any anthropologist would, how negative mental conditions are perceived and treated in other cultures. At the APA convention, the medical model of mental illness certainly reigned supreme, and that model is clearly the most widely accepted one in Western culture. But that model hasn't been accepted, or even heard of, in most non-Western cultures. How might those cultures think about, and deal with, moods and behaviors considered outside the norm?

I am also aware that the Western worldview—that is, the view prevalent in North America and Europe—constitutes a very particular, limited perspective. Billions of people around the world function within very different belief systems and economic realms. What do hunters and gatherers do with someone who slips into another reality? If we made our living growing rice or herding cows, would that eliminate depression and anxiety or make them worse? What if we had shamans and sorcerers instead of medical doctors to treat our ills? Is it possible that other cultures might suggest better approaches?

I soon realized that there might well be other models of mental illness that would be at odds with the football field of drugs at the convention. Evolution, culture, and other experiences and actions must all have dramatic effects on the biology involved in how humans feel. There must be more to the experience of mentality than brain biochemistry that can be treated only with drugs.

The Culture of Our Discontent looks beyond the medical model of mental illness and explores new paths to understanding what it means to have psychological trouble. Chapter 1 lays the groundwork with an informal look at Western culture and the impact of the medical model of mental illness. Chapter 2 considers the evolutionary model developed by a group of researchers and clinicians calling themselves evolutionary psychiatrists. This model comes from Western culture, but it is radically different from the mainstream medical model. Instead of viewing mental moods such as anxiety and depression as negative, the evolutionary model hypothesizes that these states might have had a positive role in the evolution of our species over millions of years. The biology in this context is not chemistry and cells, but the process of evolution by natural selection that produced our minds as well as our brains.

Chapter 3 comes from Western culture as well because it focuses on work conducted with nonhuman primates and what it tells us about the biological and genetic roots of our mentality. This information, too, is not the usual fare offered by psychiatrists and psychologists about the role of biology in mental states and behaviors. Instead, it underscores the interaction of biology and culture. Chapter 4 reviews groundbreaking research on the modern Western diet and how that diet affects mentality and mood. This model, which is both biological and cultural, shows that brain biochemistry is not fixed or inherited but subject to something as simple as the fats we eat. Our diet in the West, it seems, puts everyone in this culture at risk for mood disorders, a risk that might be reduced by simply changing the culture of what we eat.

Chapter 5 introduces the influence of particular cultures on individual mind-set. For decades, cultural anthropologists have observed and cataloged the thoughts, behaviors, and temperaments of cultures other than Western culture. All cultures, it seems, have their own collective personalities, belief systems, and approaches to life, and these distinctive characteristics translate into definitions of what's mentally normal or abnormal. Chapter 6 brings in the concept of culture-bound syndromes. As studies of other cultures have shown, mental illness appears in a variety of guises that are molded by the economy, ideology, and history of a culture. In other words, every culture has its own brand of craziness that makes sense within that culture. Chapter 7 explores how cultures both explain and treat mental illness relative to their own particular belief system. It may be ghosts, shamans, unhappy ancestors, or the evil eye that drive people nuts, but there are always amulets, potions, or sacrifices that can guard against or cure such problems.

The book ends on a happy note with Chapter 8. All models of mental illness, including the Western medical model, are built on hope. Sadness and despair might be the hallmarks of mental illness, but quite remarkably, every culture believes that something can always be done to bring relief.

This book on mental illness is actually a document full of hope, even if its subject is sadness and unhappiness. As long as we continue to challenge and reassess the prevailing models of mental illness, there is always the hope that we will learn something new and in turn feel happier and more balanced, and, most importantly, be able to help those who have lost their way.

CHAPTER 1

THE ROOTED SORROW

THE ACTIONS OF OTHERS ARE, IN FACT, ONLY MEANINGFUL WHEN SET against the backdrop of one's own culture. Therefore, I needed to look at mental illness in Western culture before I ventured out to look at mental illness in other places. Of course, I could have gone to Paris, or Dubrovnik, or Toronto to do my anthropological interviews on Western culture, but that seemed unnecessary, because my small hometown, Ithaca, New York, is obviously part of Western culture as well. And so instead of packing my bags, I decided to explore mental illness where I live and work.

That task was easy because Ithaca, like all American towns, is no stranger to people with mental problems. Every day I talk with one friend or another who is in an emotional crisis, or I hear about a friend of a friend who is suicidal, or has had a mental break, or is terribly, terribly sad. Walking around my town I also frequently see a few disheveled individuals mumbling and alone, clearly on a mental plane different from those around them. Mental illness and mood disturbances are all right here.

So I began my anthropological exploration in Ithaca, and decided that the best way to jump in was to interview someone who has experienced the entire range of what the Western model of mental illness has to offer. His story, I think, serves as a template for understanding the worst and best of mental illness in Western culture.

THE BETTER ANGLES OF OUR NATURE

Bill Wilson is a "babe magnet." He's a big, handsome 50-year-old with dark, wavy hair dappled with gray, and blue eyes that twinkle behind wire-rim glasses. His charm is not so much in his looks but in the way he instantly connects with people, always interested in what everyone, anyone, thinks about everything. Wilson asks a million questions, and unlike most people he really listens to the answers. Women, and men, are instinctively drawn to him because he's smart, witty, and interested in them.

Wilson's charm also has an edge, and maybe that, too, makes him irresistible. A few minutes into any conversation and it becomes clear that life is an intense business for him, and that every step forward has been hard earned.

Wilson has struggled with psychological demons all his life. These devils poke at him every day, and sometimes they push him over the edge. As he admits, "At several points in my life, if I had had access to a gun, I would be dead."

On a cold winter morning, Bill Wilson and I sit in his writing room on the third floor of the house he shares with his wife and daughter. He has agreed to tell the story of his life to illustrate what it is like to be mentally ill in this culture.

In anthropology circles, a person who agrees to talk at length about his or her own experience as a window to a culture is called an *informant*. Wilson and I agree that he will talk, and I will listen and take notes and record his experience as if I were in a far-off land listening to a stranger talk of a culture that is unknown to me. He also knows that the relationship between anthropologist and informant is sacred—these dialogues are the soul of anthropological fieldwork—and both of us know we are entering into a pact where I honor his words as truth, and he trusts me to get it right.

"Between the ages of 17 and 23, I was a major substance abuser," Wilson begins. "I was very messed up on prescription drugs—Furinol, Valium—and drinking. Finally, my parents called the cops, who brought me to a local emergency room. In the emergency room I punched an orderly, and I was taken in a straitjacket to the state mental hospital. It was a locked ward, with steel mesh on the windows. When I got there, a nurse said to me, 'If you hit anybody here, you'll be in the state hospital for the criminally insane.'"

"It scared the shit out of me. The state hospital for the criminally insane wasn't in my career plans," he says, laughing loudly and shaking his head at the memory. "I'll never forget seeing 'HP' and 'SP' next to my name on this chalkboard: homicide and suicide precautions. And I thought, *me*?"

For the next 30 days the young Bill Wilson had nothing to do but think—and observe his fellow inmates at the state mental hospital. "I had never seen a manic-depressive in full mania—ranting and raving. There were people who were so depressed they were practically catatonic, and people who twitched a lot," he recalls.

Although there was minimal treatment—some tranquilizers and weekly group therapy—that hospital stay was a wake-up call for Wilson. "I was shocked at where I was, and at the people around me," he says. "I thought, I'm screwed up, but not *this* screwed up." And that realization turned into a glimmer of hope.

"I thought of this line from Aeschylus," he recalls. "Aeschylus writes about 'the better angels of our nature,' and I kept thinking of that phrase, and knew there had to be something better in me, that I could do something and be something, and not have a life of being in and out of hospitals."

Now, 30 years later, surrounded by books stacked double on shelves clogging his writing room, Wilson says that words, classic words, were the first avenue to his salvation.

"I was always a big reader," he explains, "and in the hospital they had about 15 books in the day room. Most of them were *Reader's Digest* condensed books, but they also had *Moby Dick*, which I had read before. It was amazing to read it again while coming out of this fog of addiction."

"*Moby Dick* is about a lost young man who goes to sea," he continues, "and he meets madness in the person of Captain Ahab. And they are destroyed. After the ship is sunk and everyone is dead, Ishmael comes bobbing up on Queequeg's coffin and says, 'I alone survived.'"

Wilson puts down his cigarette and nods to himself. "I remember reading that, and it hit me that I had survived, too, and here I am, bobbing on a coffin—my coffin—and I'm far from land," he recalls. "But I'm alive. I alone survived."

About three months after Wilson was released from the hospital, a new state rehabilitation program paid for him to go to college. He loved school, read endlessly, and graduated. He then worked for a few years with deeply disturbed children, and did some writing on the side. Eventually he entered a graduate writing program, married, had a child, and established himself as a writer.

But even those successes haven't kept Bill Wilson's psychological demons at bay. Underlying the youthful alcohol and drug abuse remains a psyche in deep trouble. Sitting in his writer's room, Wilson explains that the moment of clarity in his youth was not the proverbial happy ending. Although the stay in the hospital kept Wilson from dying young, and from a life of hospitalization, it certainly wasn't a cure. Throughout his life, Bill Wilson has continued, doggedly, to deal with shifting and dangerous moods.

"My depression scares the shit out of me," he says, bringing his story to the present. "When I go down I can't get out of bed. I am paralyzed. I want to kill myself. I want to die. If I could flip a switch and die—if it would be that easy while I felt that way—I would. My depression terrifies me and I hate going there. And I have been there a lot, and it really scares me. It scares me what I think, it scares me what I could do to myself, to my family and friends."

The depression, he has learned over the years, is also complicated by periods of mania. There is no change in Wilson's body language when he describes the highs, but the words come out in a rush: "I don't sleep a lot, I work a lot, I walk around telling jokes and doing imitations—I feel really good and strong, I get a ton of work done, and I'm fun to be with, and it can last for months. But then it's followed by a deep depression."

In other words, he is often up or down, sometimes mentally soaring or plunging into a black hole. These psychological storms are, of course, a debilitating factor in his life. They have interfered with his work and his relationships, and have put his life in danger.

Unfortunately, Bill Wilson's story is not unusual.

The most recent numbers on the prevalence of all mental disorders suggest that about 18.5 percent of adults in the United States suffer from a defined mental disorder (including substance abuse) every year.[1] The World Health Organization (WHO) estimates that about 450 million people worldwide are affected by mental problems.[2] WHO also reports that mental problems are common across countries regardless of geography, culture, religion, type of government, or economic development, and that they cause "intolerable suffering" and result in "staggering economic and social costs." According to WHO, the numbers of those affected are expected to rise because of the overall aging of the human population, increasing social problems (such as poverty, unemployment, urban crowding, higher costs of living, and persistent disparities in access to education and health care), and in some countries widespread civil unrest.[3]

Bill Wilson is better off than some because he is a citizen of a country that spends more per capita on mental health research and treatment than any other country. Wilson is also a testament to the comprehensive nature of Western mental health care. Over the years he has experienced everything from an exhaustive array of psychotropic drugs and psychological tests to innumerable sessions of talk therapy.

Key to Wilson's treatment is the fact that Western culture has defined mental disorders as illnesses that rate treatments. His current diagnosis is classified by the mental health community as Bipolar Manic-Depression Type II, which usually means the depressions are extremely deep but the manic episodes are not as full blown as those he saw in the residents at the state hospital. The definition comes from the *Diagnostic and Statistical Manual, Version IV* (the DSM-IV-TR), published and updated regularly by the American Psychiatric Association.[4] Psychiatrists and others use the DSM-IV to designate a mental illness or psychological condition that qualifies for health insurance reimbursement. In addition, it is a diagnostic tool that provides a classificatory system to give mental health workers—and clients—someplace to start in understanding and treating a disorder.[5] The DSM-IV is also a cultural document because it attempts to parse human behavior into normal and abnormal based on where this particular society draws the lines. The definitions in the DSM-IV are hazy at best, and they have to be, because no two minds are identical, no two people behave alike or see the world in the same way. At the same time, it is sometimes not clear, even to those in the mental health field, who exactly is well and who is not, because there is no generally accepted definition of mental health, let alone mental illness.

The current standard of care for Bipolar II in this culture is a recipe of medication and talk therapy, and Bill Wilson has done both. He began with mood-altering medications in his early teens, and then started taking Prozac after reading *Listening to Prozac*. "When I read that book, it was all familiar, in a profound way," he explains. "Prozac worked for me; I felt like myself. The highs weren't as high and the lows weren't as low. It was miraculous. I became a believer in psychopharmacology."

But after a few years Prozac stopped working. Since then Wilson has tried a series of psychopharmacological medications with varying success. He has to take an active role in his care, monitoring his feelings and behavior and notifying his psychopharmacologist or family physician right away to avoid a drop into the black hole. "I am always wary. I am always thinking, and careful about it, and I don't want to go too up or too down," he says.

Wilson also recognizes that his particular brain chemistry is dangerously altered when he drinks alcohol, something he learned young but recently had to learn all over again, the hard way. "For years I used to tell myself that I would sooner pick up a glass of liquid Drano than I would alcohol, because Drano would be safer," he says. But that knowledge somehow evaporated one night. "A year and a half ago, after 25 years without alcohol or drugs of abuse, I started drinking again and I don't really know why." He pulls on his cigarette and recalls the circumstances.

"Late one night I couldn't sleep, and I thought, I've been sober a quarter century, I'm a good citizen, I pay my taxes and am good to my kid and wife. And I thought, I can drink," he says, his voice becoming emphatic. "There happened to be a big bottle of Wild Turkey left over from a guest's visit. I poured a big glass of it and started drinking it, and I thought, mmm, this feels good, this is nice, this is warm. And I thought, well, I'm not burning the house down and I'm not crashing the car, I'm not turning into a werewolf, so maybe I've changed, maybe I can do this."

After that night he continued to drink heavily, which led to one hospital visit and then another. He was lucky to have people around him who drew the line and confronted him when they realized what he was doing. "When I drink, all bets are off," he now acknowledges. "It's poison. Once I start, I'm not coming back. Things take over that are really terrifying."

Wilson is not alone in this reaction to alcohol. Humans, it seems, are highly vulnerable to chemicals that alter thoughts and emotions, whether for good or ill. Some people are at higher risk than others because of family history or genes, but it's impossible to predict who will become an addict until the damage is done.

Addiction to drugs and alcohol is now classified as an illness or disease by the medical community and is one of this country's most prevalent mental health problems. The National Institute on Alcohol Abuse and Alcoholism estimates that there are 14 million people—one in 13—with a drinking problem in the United States.[6] And the National Institute on Drug Abuse has published a long list of chemicals that are abused by the population in almost all age groups.[7] Abuse of alcohol or recreational drugs by a person with a mental disorder is especially dangerous.

After more than 25 years of availing himself of every sort of treatment this culture has to offer, Bill Wilson may be getting closer to understanding his mental landscape. He now takes a number of medications on a daily basis to control his mania and depression and stabilize his mood. He attends talk therapy. He goes to Alcoholics Anonymous meetings several times a week. He has recently been mulling over a piece of information about his past that was revealed during the last stay in the hospital when he underwent a battery of psychological tests.

Staying emotionally stable with medication, talk therapy, and AA, means that he just might have the mental resources to attain some peace of mind in the long term, if he can.

THE WRITTEN BRAIN

Bill Wilson believes that he owes his life, and his small measure of happiness and balance these days, to a combination of psychotropic medications and talk therapy. Both of those therapeutic tools are now so much a part of mental health treatment in Western culture that we forget that talk therapy has been around for only 100 years and that psychotropic drugs have been available for only about 50 years. Until these relatively recent developments, the treatment of mental disorders through the ages was a litany of horrors.

Not so long ago, the mentally ill were exorcised, tied up, sequestered, beaten, immersed in water, or subjected to brain surgery.[8] During the Medieval and Renaissance eras in Western culture, madness was considered a moral perversion, a perversion of the mind, or the result of satanic possession.[9] By the eighteenth century, when physicians had a better idea of how the body worked, doctors used the term "nerves" to categorize mental illness because they assumed there must be a lesion in the nervous system that caused the odd thoughts, behaviors, and moods. At that time, the medical establishment had embraced the idea that the human body was a machine and therefore mental illness must have a physical (i.e., biological) cause.[10] Following this logic, insanity was a disease that could be treated and cured, especially if the insane were removed from the stimulations of normal society and put into insane asylums. Medical historian Roy Porter also suggests that insane asylums came into fashion in Europe and North America in the nineteenth century not only as social policy for the mentally ill, but also as a way for the growing mental health system to turn a profit.[11]

According to Porter, psychiatric treatment over the past 200 years in Western culture has always been oriented toward restoring balance to the mind and brain through outside agents. These agents have included a broad array of both naturally derived and, more recently, chemically manufactured substances. One of the oldest and most popular of these brain-altering agents is alcohol, which is a depressant that reduces anxiety. Alcohol has, of course, been around ever since people discovered fermented fruits and vegetables (and even some nonhuman primates enjoy becoming intoxicated). Humans also have a long history of smoking, chewing, swallowing, or shooting up various other substances for mental alteration. The use of natural chemicals to blunt mental pain can be traced to the extraction of opium from poppy flowers to create sedatives and painkillers. Laudanum, a recipe of alcohol and opium, was used for centuries for all kinds of physical and mental complaints, and was an acceptable sleeping potion until the latter part of the nineteenth century.[12] But the way modern psychotropic drugs are administered, used, and accepted today began with the major breakthroughs in the early 1950s that accompanied the development of antipsychotic medications.[13]

Psychiatrists had long been looking for some way to help schizophrenics quell their inner voices and experience a calmer reality. In the late 1940s a French military doctor noted that his patients lost their anxiety when given the drug phenothiazine, initially an anti-histamine. Researchers quickly recognized the potential for this class of drugs on mental patients, and by the early 1950s a new version of the drug, called chlorpromazine (better known as Thorazine), was promoted as a miracle drug that reduced the major symptoms of schizophrenia. Thorazine and other antipsychotic drugs revolutionized the treatment of serious psychiatric illness and psychiatric practice in general; people who had been shut away in asylums were now able to function in the real world. The rapidly growing use of antipsychotics and their apparent success with serious mental illness soon suggested to the culture at large that mental disorders were easily defined conditions curable with medication, like infectious diseases.[14]

Shortly after the discovery of antipsychotics, researchers developed antidepressants.[15] Among the many generations of antidepressants, those called SSRIs, or selective serotonin reuptake inhibitors, are the most commonly prescribed today because they have few side effects and are not addictive. But they have limitations: SSRIs do not work for everyone, and there seems to be a very high placebo effect.[16] Trials of antidepressants have shown that at least 40 percent or more of the subjects said they felt better even when they were taking a placebo and not the antidepressant. It appears that the interaction with a caring staff and the simple thought that life might actually improve with chemicals contribute to a positive mind-altering experience.

The introduction of psychotropic medications has revolutionized the health care industry and deeply affected the way people in Western culture view both mental disorders and treatment. At this point, therapeutic options for the mentally ill in Western culture always involve the possibility of medication (including the decision not to prescribe). Beyond helping those with serious mental illnesses, pills are advertised to cure or reduce the effects of more common problems, such as depression, mania, premenstrual syndrome, obsessions, compulsions, bulimia, anorexia, menopause, anger, vague feelings of sadness, and anxiety. In addition to these ailments, historian Porter observes that the development of designer drugs for mental conditions has coincided with a flowering of newly coined mental disorders, such as Post Traumatic Stress Syndrome and repressed memory, all of which are candidates for medication.[17]

With this trend in our culture in which medication for mental disorders is the norm, psychiatrists are now more valued for their medical training and knowledge of pharmacology than for their psychoanalytical skills. In fact, psychologists and social workers—not psychiatrists—now do most of the talk therapy. But they aren't licensed to write prescriptions and must refer clients to medical doctors for medications. This hierarchy enforces the idea in Western culture that mental disorders are primarily biological, and therefore medical, phenomena.

And so it is to a psychiatrist that I turned to understand the Western medical model of mental illness.

DR. HOWARD FEINSTEIN, A PSYCHIATRIST in Ithaca, is well placed to address the issue of recent shifts in the Western approach to mental illness because his own career shifted following the introduction of psychotropic drugs to mainstream psychiatry. Feinstein is a classically trained psychoanalyst, and he looks like one. Thin, tall, with a gray beard—he would look perfect sitting behind a couch taking notes and nodding while someone talked. Feinstein attended medical school in Boston and then completed a residency at the Massachusetts Mental Health Center. In the 1980s he became interested in what the new drugs were doing for the mentally ill and set about retraining himself as a psychopharmacologist, that is, a specialist in the use of drugs to aid in the treatment of mental illness. In the process,

Feinstein became a convert to medication as integral to mental health, and he is a real supporter of the pharmaceutical industry. "They have put miraculous tools in my hands and I have seen miracles in my office every day—and I have been at this a long time," he says.

Feinstein's conversion, he explains, is practical: "Psychoanalysis is very labor intensive, regardless of whether or not it is effective. The new technologies are labor efficient and they are directly effective on the disease symptoms." Feinstein also acknowledges that prescribing medication and seeing its effect changed his view of mental illness. "I used to believe that people can think their way out of these problems," he explains. "I don't believe that anymore. I'm not ambivalent about medication at all; I'm very enthusiastic."

Feinstein adheres to the idea of the body as a biological machine. "When it comes to making a decision about which medicine and how much for mental problems, it's a medical model, a mechanical model," he explains. "It's a complex model, fully recognizing that you are dealing with a person, but it is 'man-is-a-machine.' For example, no matter what the cause, once the depressive process begins, it's all set. You get the same mix of symptoms of depression, and people will respond to the same medications whether they are depressed because they have cancer, or they are depressed because they have a broken relationship, or they are depressed because the switch just flips on its own."

Pharmacological intervention, he believes, is essential. "You can neutralize their illness," he says. "The statistics in psychiatry are much better than they are for heart disease in terms of cure rates, but that depends on people getting the proper diagnostician, getting the proper medicine, and sticking with it."

According to Feinstein, these drugs aren't used often enough. "There is great cultural and individual resistance to medication for mental and behavioral problems," he insists. "In fact, many of the things in this culture are hostile to treatment. There is disbelief, and then there is 'pull yourself up by your own bootstraps.' Also, strains in the culture, which are very old, emphasize solitary independence as an unqualified good. Add to that in our particular community the current fad of naturalism: If it's vegetables, it's good; if it's manufactured, it's bad."

From Feinstein's point of view, this resistance gets in the way of helping people. "I spend a lot of my day trying to get people to accept treatment," he says.

Nonetheless, in only two decades psychotropic medication has become a generally accepted part of mental health treatment in Western society; in that sense, such medications are now a fixture of the culture.[18]

No one knows the long-term effects of a population altered by mental medication.[19] For example, how might these drugs change life experiences or even permanently alter the delicate biological mechanisms that govern the brain? The mind is a sponge that soaks up stimuli and translates those stimuli into a million thoughts; changing the biochemical soup of the brain makes for different thoughts and feelings, and thus different experiences and perceptions.

Some practitioners are indeed wary of the use of medications, especially for cases in which a depression appears to be temporary. I shared with Feinstein a personal anecdote of my own encounter with the mental health profession for the treatment of a situational depression. About 15 years ago, everything in my life went wrong, and I became very depressed. I approached a psychiatrist and brought up the possibility of antidepressants, but he refused, explaining that my problems were situational and that when things were better in my life, I wouldn't be depressed any longer. It took about three years for me to bounce back, and as I built a new life, my mood did indeed improve.

When I related this story to Feinstein, he nodded, but suggested that if I had been on antidepressants, it might not have taken so long for me to make my life better, and I might not have been so unhappy during the time it took to fix my problems. "That's probably true," I agreed, "but I am extremely proud of what I did, and I take exclusive credit for making my life better. I also learned a lot about myself during that time."

Feinstein nodded again, acknowledging the efforts and rewards that come from such a process, but he stuck to his point. "But you might not have spent all those years unhappy if you had been on medication," he reiterates.

THE TALKING CURE

Although psychiatrist Howard Feinstein is all for medication, he also knows that effective and responsible treatment of mental illness is not just about writing a prescription. "You have to help people overcome resistance to medications, deal with their changing sense of self, and engage with family members," he explains. "Talk therapy is essential."

Easing the mind through talk has a long history in Western culture. In the late 1700s an approach called moral therapy emerged in Italy, France, and the United States.[20] Its practitioners believed that the insane could be cured if their moral reasoning, their thought processes, could be redirected. Therefore, talk became part of the cure, either in addition to or instead of the typically prescribed rest, isolation, restraint, or other physical treatments.

At the beginning of the twentieth century, the writings of Sigmund Freud captured the attention of the psychiatric profession and talk therapy soon became entrenched. Freud knew that talking could be used only with the mildest cases, and that schizophrenics and others with serious mental illnesses required more than therapeutic conversation.[21] For those with common neuroses, such as anxiety and depression, Freud believed that talking about oneself could be curative. According to Freud, much of our behavior is dictated by unconscious thoughts and motivations; in particular, unresolved sexual conflicts stored deep within the mind are the source of many neuroses. He advocated a process called psychoanalysis, in which the patient gains insight through the random exploration of thoughts and feelings (called free association). Those insights will eventually release the patient from the influence of whatever was hidden deep in the unconscious. The therapist acts as a largely silent guide in this exploration.

Not everyone accepted the premise of psychoanalysis. In the 1920s psychologists J. B. Watson and B. F. Skinner challenged Freud's approach by suggesting that environment molded a person, not the unconscious.[22] According to the Social Learning, or Behaviorism, theories developed by Skinner, for example, people can replace negative behavior patterns with healthier ones if they have the right guidance. The focus of this approach is not on introspection but on action and change.

The split between the psychoanalysis school of therapy and the behaviorism approach is evident today in the practice of both types of talk therapy in Western culture. A person who works with a Freudian therapist (i.e., one who is psychoanalytically oriented) will participate in sessions that focus on self-reflection. Presumably, insights about the client's past will shed light on current behaviors and feelings. Alternatively, a person may choose a cognitive-behavioral therapist more interested in helping the client change unhealthy or troubling behavior patterns. Cognitive-behavioral therapy has its roots in Skinner and Watson's view that change is possible given the right environment.

No one has been able to prove scientifically that either of these options, or any of their subtypes that have flowered in the past 40 years, really help, or that one is more effective than another.[23] Few studies on the efficacy of various kinds of talk therapy are able to meet the criteria for the kinds of double-blind clinical trials that clearly point to a well-defined or measurable effect. In addition, it's difficult to determine whether a client's improvement is due specifically to a particular technique or some other factor. It may be that the client finally has a sympathetic ear, or perhaps that the passage of time has brought some relief.

Yet in Western culture, talk therapy has become standard practice for the treatment of mood disorders, substance abuse, behavioral issues, and interpersonal conflicts. I felt it was important, therefore, to include a therapist as an anthropological informant on mental illness in Western culture. Can a depressed or anxious or compulsive person really become more functional just by talking? Why would someone be drawn to a career designed to listen to other people's problems?

Looking for a therapist to interview proved harder than I expected simply because there are so many kinds of therapists. Those seeking help can call their family physician, their minister, a paid therapist, or a suicide hotline. They may choose to see a psychiatrist, a psychologist, someone with a master's degree in clinical social work, or a certified family therapist. The Yellow Pages for my town alone list 3 psychiatrists, 28 licensed psychologists, and 13 psychotherapists who have counseling degrees in either social work or some sort of talk therapy. Additional mental health workers with various certificates, degrees, and training are associated with social services agencies. And there are therapists and counselors affiliated with both the university and the college in my town, as well as psychologists and counselors in the school district. From the myriad choices available, I contacted Tamara Loomis, a behavioral therapist, who was willing to sit down and talk about her work.

Loomis grew up in Putney, Vermont, a picturesque village with a town square bordered by a flowing stream. Loomis has loads of curly strawberry blonde hair and big brown eyes, and it's easy to imagine her background as a kid in a country village. As a child and teenager in Putney, she can't remember a single person identified as "crazy" or even depressed. Of course, she grew up in an era when someone with a mental illness was either sent away or simply not discussed. Yet she has spent her adult years trying to help people crippled by their own feelings.

Loomis majored in both psychology and business as an undergraduate and thought she would go into personnel work, but the city of Ithaca, to which she has a deep commitment, doesn't have many jobs in that area. So she decided to focus on psychology, and that turned out to be an education in itself.

Loomis started working in a group home for seriously disabled adults. "Some people were schizophrenic, some bipolar," she explains. "We didn't do [talk] therapy with them. We helped them with activities of daily living, and we administered medications." She admits that the job was a challenge. "At first I was nervous," she says. "But once I got there, the clients were so gentle, perhaps because of the developmental issues."

Loomis worked for two years at the group home, and then enrolled in the master's of social work program at Boston University, which forced her into situations not seen in Putney or Ithaca.

"During my first internship, I worked with homeless veterans," she says. "A lot of them were mentally ill, many suffering from Post Traumatic Stress Syndrome, and they chose to be homeless. My job was to offer them services from the Veterans Administration, but I usually only saw them once and then they disappeared."

In the second year of her internship, Loomis worked in an office at a social services agency with single adults, children, and families. When she completed her master's degree, she returned to Ithaca to work at Family and Children's Services, a nonprofit mental health and social services agency.

Many of her clients at Family and Children's had to adhere to the constraints of their employers' mental health benefits. "Employers will pay for only six to eight sessions," Loomis explains, "so you can't do the tell-your-life-story kind of therapy—you only have eight sessions."

As a result, she was drawn to an approach called Brief Therapy, a highly interactive talk therapy aimed at getting people to solve their problems rather than ruminate on them. "One of the truisms of talk therapy is that long-term therapists can have the same clients for five years and not have to do much different," she explains. "My colleagues and I have a constant turnover, but we feel that this kind of brief, proactive therapy is needed; I've seen that it works after eight sessions—that people don't need to come anymore."

This kind of dynamic therapy resonates personally for her as well. "I just wasn't good at sitting back and saying 'uh-huh, uh-huh,'" Loomis says, nodding her head to demonstrate. "I was bored with long-term sessions, and I felt people became incredibly dependent on it. It was seen more as 'Well, I can meet my friend/therapist once a week and I don't have to do anything else.' It becomes a major crutch." In contrast, her role in Brief Therapy is to suggest that a client try specific steps toward improvement, and if one doesn't work, together they create a new plan. As Loomis puts it, "There is always constant movement."

In 1998 Loomis and a colleague left Family and Children's to offer only Brief Therapy; today their practice is thriving. "I have seen more clients in the past six years than most therapists see in a lifetime," she says. "And I have seen all sorts of disorders and problems. Most people come about 15 or 20 times, but I do have some long-term clients; they come infrequently, say, once every three months. Some people protest that their issues need more time and that they want to go slower. Usually it's because they don't want to change. Our motto is, 'No more therapy than necessary.'"

Brief Therapy is not suitable for all mental problems, so Loomis and her colleagues do not see people with serious mental illnesses, such as schizophrenia and bipolar disorder. That still leaves a range of troubles that they help clients address. "Over and over, people come in because the husband and wife are not getting along. The three things couples fight about are sex, money, and housework—and if the last two can be worked out, they would have sex all the time," she says with a smile. Loomis also has extensive experience with a more somber concern: "Sexual abuse is also a major recurring issue." She cites several stories in which sexual abuse in childhood deeply damaged her clients. Even in these cases, Loomis works to help the clients make changes in their lives rather than focusing on the history of the damage.

THERE ARE NO STATISTICS ON visits to talk therapists in Western culture; the numbers would be impossible to get or verify given the variety of therapists and venues, but clearly our notion of mental health treatment now typically involves some sort of interpersonal dialogue. But there is an inherent conflict in this view. In Western culture, mental illness is most often considered a disease or illness and therefore suitable for medication, yet the belief also persists that talk is integral to treatment.

Informal surveys and anecdotal evidence suggest that most people have seen a therapist, or are seeing one, or expect to see one. For example, a rudimentary calculation might take just half the number of clients that Tamara Loomis sees and multiply that by the number of talk therapists in town. The result would indicate that every citizen in this town has had a recent or anticipated appointment with a talk therapist. What does this phenomenon say about Western society and mental illness? What is the role of society in the care and treatment of the mentally ill?

THE SYSTEM

The Tompkins County Mental Health Building is one block over from the center of downtown Ithaca. Six stories high, it anchors the eastern part of downtown, looming over the public library. When the building went up, some wondered, "Why does it have to be so big? What in the world do they do in there?" In the context of this anthropological survey, it made sense to go in and find out.

I knocked on the door of Robert De Luca, a clinical social worker and the commissioner for Tompkins County Mental Health. De Luca is 50, with graying brown hair, a sparse brown beard, and the calm voice and manner of someone who keeps a healthy perspective on what he can't change. He also seems unfazed by the complexity of bureaucracy that keeps the public mental health system afloat in these lean financial times. "Every state in the union has different structures for dealing with mental health," he explains. "Our county in particu-

lar is a large service provider, which is not true of every county." The
Tompkins County Mental Health Building is tall because it has to be.

The history of local public assistance for mental problems in
Western culture can be charted in the rise and fall of asylums and
mental hospitals in Europe and the United States over the past 200
years.[24] In the 1800s there were fewer than 5,000 patients in mental
institutions in the United States, but by 1904 more than 150,000
people were housed in such facilities, and by 1950 the number had
risen to half a million.[25] These figures don't indicate that more citizens were troubled but that sending them away had become both
social policy and culturally correct.

The past 30 years have witnessed the reversal of the asylum movement. With the development of psychoactive drugs, states have been
closing mental facilities and developing community-based programs
for every level of mental illness; culturally, it's no longer "right" to
send loved ones off to a state hospital unless they can't be managed at
or close to home. "The question always is," says De Luca, "How much
will society tolerate?"

New York state is a good barometer of Western culture's changes
in mental illness treatment because it's one of the few states with a
constitutional mandate to care for the mentally ill, no matter their
ability to pay. De Luca sketches a brief history: "Around the turn of
the century, this state built massive psychiatric centers, big campuses.
In New York state in 1955, we had 95,000 people in state-run psychiatric centers." Even when the patients were deinstitutionalized, the
state was still in charge of their care. "At one time, county mental
health was truly dealing with deinstitutionalized people who had
spent many, many years inside psychiatric centers," explains De Luca.
"As a result, some of our services were set up to cope with the consequences of spending many years in a confined environment. The need
for day treatment for these particular clients has been reduced greatly
because they have aged or passed away. For others we try to manage
illnesses with drugs and support services."

Other social movements have also influenced policy and treatment. "The concept of 'recovery' is promulgated these days," De Luca adds. "The consumer movement has also had a big effect, as it should."

Given this history, Tompkins County Mental Health plays a significant role in the management of mental illness in this particular microcosm of Western culture. For those seeking help, there are three psychiatrists, two nurse practitioners, and various psychiatric social workers, mental health nurses, and case managers available to respond. They evaluate the client's condition, and may prescribe medication, conduct talk therapy, refer appropriate cases to the substance abuse clinic, or help with day-to-day management of food, housing, and money. Emergency cases can be seen right away, and house calls are a possibility too.

The array of options accounts for New York state's high taxes. This state spends the highest amount per capita ($175.97) on mental health care in the United States.[26] "It is state law that the local government unit must plan and provide for those who cannot provide for themselves," De Luca explains. "Our big job is to provide a comprehensive system of care and make sure the system exists regardless of ability to pay."

The challenges include more than clients' ability to pay for mental health care. Outside the Tompkins County Mental Health Building was a group of smokers, and they bore the marks of a low-income background—bad or missing teeth, shabby clothing, and the physical ravages of poor nutrition, inadequate medical care, and hard lives. One might assume that their poverty led them to utilize the resources of the Tompkins County Mental Health program. The picture is more complex than that, however.

"The sociological point of view is that poverty increases all stressors, so if you have a tendency toward a particular disorder or illness, it increases the chances that you will be ill," De Luca says. "The mirror point of view is that mental illness is impoverishing. Let's say you go to college and you are a premed student and then in your third year of medical school you have a schizophrenic break. You have gone from someone who could earn $500,000 a year to someone whose basic means of support will be the Social Security system."

De Luca's point is that the group outside is probably a mix of those who started out poor and became mentally ill, and those who fell into poverty because of their mental illness.

When asked about the impact of poverty on mental illness, he answers quickly and softly: "Power. It's all about power."

In Western culture, money is power, and so poverty frequently means less power and fewer options. People with money can seek out any treatment; people living below the poverty line have to rely on public assistance, which usually means increased hardship in states that spend little on mental health. Nonetheless, in Western culture we still believe that a civil society is supposed to help those who are unable to provide for themselves, and it seems that we often do.

According to De Luca, there are no easy answers for why some people need mental health help and why others don't, or whether medication or talk therapy is more effective. Instead, he acknowledges that doctors and patients alike have been frustrated by mental illness for a very long time.

"Do you like Shakespeare?" he asks shyly at the close of our interview. "Macbeth goes to the doctor in anger and desperation because Lady Macbeth has lost it. Keep in mind that Macbeth is going to a doctor to cure his wife of mental illness." De Luca's point is that even back then people were turning to medical doctors in search of a cure for mental illness and not always getting an answer.

The commissioner of Tompkins County Mental Health, who has spent his life providing mental health treatment to the masses, then recites, with perfect diction and cadence, the frustration voiced centuries ago by Macbeth and shared today by those who still struggle with mental illness:

> Canst thou not minister to a mind diseased,
> Pluck from the memory a rooted sorrow,

Raze out the written trouble of the brain
And with some sweet oblivious antidote
Cleanse the stuff'd bosom of that perilous stuff
Which weighs upon the heart?

THE CULTURE OF MENTAL ILLNESS

I am sitting in a nondescript office building in Bethesda, Maryland, surrounded by the suburban sprawl that has consumed the greater Washington, D.C., area. I lived in Bethesda as a child, but I barely recognize the place anymore. The suburbia of my youth has changed in myriad ways, from the eight-lane roads crowded with minivans and SUVs and numerous new strip malls full of national chain stores to the tightly packed apartment complexes that have spread over the landscape.

The demographics of the area have also changed. Whereas the population of the 1950s and 1960s was predominantly white, today there are immigrants from all over the world, people of different races, languages, and customs.

The United States and other developed nations have become cultural melting pots, and so what we refer to as Western culture is, in fact, a dynamic amalgam of cultures—as well as part of an increasingly global culture that is rapidly connecting everyone on Earth. I have come to Bethesda to meet Dr. Javier Escobar, senior advisor to the director of the National Institute of Mental Health (NIMH) because I need a broader view of mental health in Western culture. Escobar was brought to Washington from the Department of Psychiatry at the University of Medicine and Dentistry of New Jersey, Robert Wood Johnson Medical School, to help develop a global office for mental health within NIMH. He seems like a good source for putting American views within the larger context of Western culture, and then moving beyond the West into other cultures.

"'Global' is a fancy and fashionable word that everyone is using now," Escobar says as we begin. "But it also means that we need to think in international terms about mental health, and if we don't, we may be missing a good part of the picture." It is quickly obvious why NIMH chose Escobar to be an ambassador for mental health. He is well qualified as a psychiatrist, has the manner of a gentleman and diplomat, and is engagingly energetic. As an immigrant (he was raised and educated in Colombia), Escobar brings an outsider's perspective to the U.S. role in studies of mental illness.

In fact, he doesn't even believe that the United States is the best role model for the treatment of mental health. "I believe that we should go international first and see what the rest of the world tells us, and learn from mental disorders worldwide," he says. "In this country we have so many people from so many places in the world, and they have the tendency to retain some of their basic cultural values. I think anything we can learn from, say, Latin American, African, or Asian countries will help us understand what is going on in America."

Escobar goes on to point out that there is decent, even better, mental health care in other countries. "There are many countries that have developed high-quality mental health systems," he claims. "They don't have the technologies we have, but the community services and outreach are much better." Escobar cites the example of research on schizophrenia across cultures: Although schizophrenia is found at about the same rate in all countries, the outcome—or prognosis—of the condition varies widely. In fact, people diagnosed as schizophrenic do much better in developing countries than in developed countries, regardless of treatment.[27]

"Many cultures that are less individualistic—more communal— are much better at maintaining and dealing with mental health," Escobar says. There is also evidence that current mental health treatments in developed countries—such as a ready supply of psychotropic drugs—can make conditions worse for some patients. "For example, in the early years of antipsychotic medication, people were given massive doses. We paralyzed patients and they had no social function," he explains. "When you look across cultures today, there is a difference in the way practitioners use medications. For example, in Latin American countries, even for depression, clinicians tend to use much lower doses than we use in the U.S."

The United States can benefit from other countries for research purposes, he says. For example, an effective survey of the genetic component of mental illness requires large family networks that live in close proximity, and these are few and far between in developed countries but common in other cultures. Other countries can also serve as collaborators in addressing very broad issues of mental illness in ways that simply can't be done in a single culture or country, even if it is a melting pot.

In addition to these practical arguments, Escobar observes that U.S. mental health care can be enhanced through global collaboration because Western society seems oblivious to the biasing power of culture. Even in these politically correct times, race and ethnicity have a persistent—and generally negative—effect on the diagnosis, treatment, and outcome of the mentally ill in Western culture.

Escobar cites an NIMH study that evaluated the symptoms and subsequent diagnoses of over 19,000 clients seen for the first time in the Institute's 22 mental health clinics. "We discovered that regardless of the presentation of symptoms, African Americans still tended to be diagnosed as psychotic. And this diagnosis is unfortunately the most stigmatizing, and will most often lead to the use of medication," Escobar says. At the same time, he continues, Latinos tended to be diagnosed as depressed, regardless of their symptoms, even when they presented with psychotic symptoms such as hallucinations.[28]

Escobar suspects that language is partially at fault. "Many of these people are not well acculturated, only speak one language, and the clinicians spoke to them through interpreters," he says. "They saw the patient subdued and dejected, and they could not communicate, so depression [was] the diagnosis." What distresses Escobar the most is that the study was conducted by researchers in his own training program, which includes sensitivity training on issues of race and ethnicity. "Imagine what it is like in places in the United States where such biases have not even been addressed," he says, the frustration clear in his voice.

Language is not the only challenge for immigrants in the United States. Psychiatric researcher William Vega and others have shown that the children of Mexican immigrants in the United States are twice as likely to have psychiatric disorders as their parents, and they are four times as likely to abuse drugs and alcohol.[29] The same is true for the offspring of European immigrants to the United States, and for those of all immigrants in Canada. Apparently, coming to a developed country can pay off economically, but immigration also puts the next generation at high risk for mental and behavioral problems. What goes wrong from one generation to the next?

Escobar offers an explanation: "Perhaps you are escaping a bad situation and you are willing to undergo some trials and tribulations. But if you are born in a country, you feel entitled," he says. "It also has to do with discrimination. Second- and third-generation immigrants identify themselves as minorities, which immigrants rarely do, and they are marginalized and stay in certain neighborhoods. It's very hard for some people to make it in the American system, and yet they feel entitled because they were born here, are American citizens. Their parents may have lived in the same neighborhood, but their expectations are much lower."

Some subcultures also do better than others over time. "I strongly suspect that Mexican immigrants have an advantage because they still have a 'protective' or 'buffering' effect of traditional culture," Escobar has written. "Traditionally, Hispanic families have been described as close-knit, extended family networks that offer a great deal of support. Retention of cultural traditions may also contribute to healthier habits (e.g., better eating and less drug use) that may lead to better health and mental health outcomes."[30]

The adoption of new cultural mores and traditions may even trigger genetic risks that might otherwise remain hidden. William Vega believes that genetic predisposition for certain mental conditions may manifest in this culture because of differences in what is socially sanctioned. He claims, for example, that there is an increase in alcoholism among women because drinking by women is tolerated in this culture, whereas in a traditional society it is not. Vega and others have pointed to the rise of obesity and diabetes in Western culture as analogous to mental illness—the predisposing genes are there and are activated by changes in diet and other behaviors that result in new health risks and problems.

Culture is not only about race or ethnicity. "Not long ago, when the United States was in the clutches of psychoanalysis, most people with severe mental illness were diagnosed as schizophrenic," Escobar describes. "But when they went to England, the same people were diagnosed as manic-depressive. The 'joke' was, to be cured of schizophrenia, you just had to fly to England."[31]

The idea that culture alone might influence how physicians diagnose and treat their patients is not new. In 1988 medical journalist Lynne Payer published a startling analysis comparing various health conditions and their treatments by physicians in the United States, Germany, France, and Great Britain.[32] Even among these cultures that have much in common, including high socioeconomic status, there are startling differences in the way physicians view diseases, the prevailing standards of care, and the use of both tests and medications. For example, Payer writes, French doctors tend to attribute many complaints to the liver, German doctors believe low blood pressure is dangerous, British doctors usually take a wait-and-see attitude, and American doctors favor aggressive testing and surgery. Yet the life expectancy is about the same for all these countries.

"American culture, we love it," Escobar says. But he also sees the downside to becoming citizens of this country. "We ought to put a sign at the Mexican-American border, on the U.S. side, that reads, 'Entering this country can be hazardous to your mental health.'"

THE EXPERIENCES AND OBSERVATIONS of Bill Wilson, Howard Feinstein, Tamara Loomis, Robert De Luca, and Javier Escobar indicate that Western culture is just as vulnerable to trends in mental illness treatment as any other culture. Experts in the United States and other Western countries may think of themselves as sophisticated and well equipped to deal with challenges such as mental disorders, but we are as much governed by history, tradition, folklore, and opinion as any less economically developed culture or country.

That's because the human mind, no matter the cultural context, is a fragile instrument, subject to anxiety, depression, psychosis, and a host of other problems that make us unhappy, fearful, worried, and insane.

Clearly, humans are a species with inherent mental problems. Why are we so cursed?

CHAPTER 2

THE EVOLUTION OF THE MIND

"A WEALTHY 45-YEAR-OLD WOMAN IS SUDDENLY STRUCK WITH ANXIETY that she has never had before. Her husband is a lawyer, her kids are doing fine, and she has no idea why she's so anxious. But she wants me to write her a prescription to fix her life."

Psychiatrist Randolph Nesse of the University of Michigan tells me about this patient as if I already know her. And I do. I can readily think of several friends with seemingly great lives—good marriages, decent jobs, lovely homes, happy kids—and yet they are on antidepressants. Unlike some of his colleagues who judge such patients as worried for no good reason, Nesse is particularly intrigued by their stories because he believes that worry in the face of apparent happiness is not a pathology but an appropriate evolutionary strategy.

Nesse and a growing number of psychiatrists believe we are ignoring an essential part of the explanation for mental disorders and I have come to Michigan to hear his perspective. Darwinian or evolutionary psychiatrists like Nesse are convinced that evolutionary theory, rather than a medical or psychological model, is key to understanding mental illness and mood disorders. They are also convinced that current psychiatric practice might be transformed if psychiatrists took an evolutionary or a Darwinian perspective of their patients before pulling out the prescription pad.

Nesse continues: "I start to talk to her and she mentions her husband's golf buddies and how her husband isn't like them, that he hasn't picked up with a new, young wife. Golf buddies? One of them has a new young wife? No, the woman tells me, four of them have new wives. And what about you, I ask? Oh, my husband says he'll never leave me."

"Well, what does she think about that?" he asks rhetorically. "She's very anxious. This woman has everything going for her. And her unconscious is absolutely accurately guiding her. Her whole life adaptation is at risk, and there's not that much she can do about it."

The woman seeking mental health care, according to Nesse's evolutionarily informed diagnosis, is appropriately anxious. Her anxiety is not a figment of her imagination, but it is also not a disease, nor a product of a bad childhood. It's a case of the adapted human brain correctly assessing its situation.

In other words, her brain evolved to make her worry.

EVOLUTION AND THE MIND

The evolution of the human brain occurred over 4 million years during which the brains of our ancient ancestors grew from the size of a chimpanzee brain, about 450 cubic centimeters (roughly the size of a softball), to today's size of 1,300 cubic centimeters.[1] Although this increase was gradual over those 4 million years, there was a major upward leap in brain size 1.5 million years ago that no one can really explain.[2]

Anthropologists believe that no matter the details, something essential and dramatic must have favored those with more brain cells. It may be that humans were beginning to rely more on tools; there were primitive tools before this time, but the sudden steep growth in brain size seems to coincide with an increase in the type and quality of tools found with those human fossils. Some scholars have suggested that something more social was at work.[3] Perhaps the complex social groups so typical of our order—the primates—favored better brains in humans. With bigger brains, our kind could more easily keep track of complex interpersonal interactions, an ability that must have been advantageous for a variety of reasons. For example, enhanced social awareness could be useful to form alliances for mutual protection, for keeping track of favors, and for pooling resources. In any case, the evolutionary pressure to increase the brain's size and complexity must have been intense for our ancient ancestors, and the brainiest ones clearly passed along the most genes. Today we have the largest and most convoluted brain relative to body size of any animal on the planet.[4]

We presume that having a big brain is a good thing. Brain tissue is expensive to maintain, in terms of the energy, oxygen, and blood flow required for even basic functioning. Surely, natural selection would not have opted for such a costly body part if it weren't ultimately useful. In fact, the brain is the most costly organ in the body: while it represents only 2 percent of body weight, the brain uses about 20 percent of our daily calories.[5] What we get for all that caloric consumption is the site for consciousness, intelligence, decision making, memory, speculation, language, and behavior.

That large brain package also includes the more ephemeral mental capacities, such as emotion, feeling, and mood. No one knows whether these psychological abilities were simply extras that came with a big puzzle-solving brain, or whether they were also adaptive. Imagine that human brain size was selected not for intelligently solving puzzles, as most anthropologists believe, but for emotional dexterity. For all we know, our ancestors may have had better reproductive success not because they were smart, but because they were emotionally sensitive, dynamically moody, and ridden with anxiety.

In other words, it makes sense to hold up an evolutionary lens to mental disorders because they might not be disorders at all, but adaptations.[6]

If that view is correct, what are now considered mental illnesses might in fact represent past adaptations that are currently mismatched with today's cultural and social pressures. Or they may still be adaptive when the larger view of the client's life is taken into account.[7]

DARWINIAN PSYCHIATRY

The idea that evolution might be behind many mental disorders is not new, although it has a spotty history. In the middle 1800s Charles Darwin was keenly interested in the mentally ill. He assumed that the insane had been robbed of reason and their behavior simply reflected an underlying brutal nature that must be shared by all humankind.[8] Darwin believed that the insane would therefore be good study subjects for understanding the role of emotions in human behavior. Darwin's view, of course, is contrary to any of today's models of mental illness, but he did have an evolutionist's eye on what it meant, in the greater sense, for the mind to malfunction. He at least considered the mind a product of evolutionary forces and paved the way for making evolutionary hypotheses for various mental states.

Many years later both Sigmund Freud and Carl Jung attempted to incorporate evolutionary theory into their early writings on mental disorders.[9] But their evolutionary bent was soon discarded and replaced by individual psychoanalysis, where evolution has no purchase. For many decades only a few psychiatrists, such as Aubrey Lewis in the 1930s, felt that evolution had anything to do with mentality.[10] But by the 1980s evolutionary thinking had permeated all branches of the study of animal and human behavior, so it made sense to test evolutionary theory on thoughts and mood as well.[11]

Since the 1990s, an entire industry of evolutionary psychologists has been looking at many aspects of human mentality, but this group has been more interested in how humans think or strategize than in the concepts of mental health or mental disorders.[12] In other words, they are applying evolutionary theory to what is considered normal thought. In contrast, the less visible cadre of researchers called evolutionary psychiatrists or Darwinian psychiatrists is also interested in how evolution molds the mind, but they focus solely on the evolution of mental health and illness.

The Darwinian psychiatrists have become proactive in recent years because they are decidedly uneasy with the turn that psychiatry has taken in Western culture, especially in America. They see an escalating trend toward drugs and a turning way from more thoughtful ways of approaching the troubled mind, and they wonder how any psychiatrist can make judgments about what is normal or abnormal behavior without a general theory of human behavior.[13] Darwinian psychiatrists also question how any therapist can prescribe medication without considering the possibility that evolution might have a significant role in producing human mentality. Just because we call a mental mood bad now, that doesn't mean it was always bad. Most of all, Darwinian psychiatrists see mainstream psychiatrists as inflexible and narrow in their approach and unwilling to consider explanations other than the standard medical model.

"When depression is caused by a situation, it might mean you should be on antidepressants, just as you might take an aspirin after you hit your thumb with a hammer," says Nesse. "But on the other hand, to automatically give everyone antidepressants because of the idea that depression is a brain disorder is thinking at a 10-year-old's level."

Instead, Darwinian psychiatrists suggest that psychiatric practice should be informed by the idea that the human mind, like the human body, was shaped by natural selection—that human mentality evolved in certain ways to cope with certain situations—and we need to know the evolutionary context of our mental state before we assume every problem is due to pathology.

More unnerving, the Darwinian psychiatrists suggest that we are completely wrong in assuming psychological peace is akin to happiness. Under an evolutionary rubric, the point is not happiness, but individual reproductive success. "The human mind was not adapted to be happy," writes Nesse. In fact, our minds might have evolved for more general challenges than being mentally balanced. "Evolutionary theory suggests that humans are adapted for maximizing reproductive success, which may involve speedy and excessive responses, rather than cooling, logical, calculated ones," explains Darwinian psychiatrist Paul Gilbert.[14]

Taking an evolutionary view, then, can tip standards of care on end. Instead of trying to make everyone "happy," there might be some other ways to help those in the clutches of mental disorders regain a sense of balance.

Nesse, in particular, considers negative feelings more as defenses than disorders, in the way that nausea and vomiting are defenses against a stomach bug or a cough is a defense against pneumonia.[15] He points out that while the symptom is useless, the physiological defenses are the way the body copes with infection. In the same way, he feels the mind produces defenses against adversity. As Nesse writes, "Natural selection has molded each kind of bad feeling to help protect against a specific threat."[16] Negative moods, in this scheme, operate like pain. Just as physical pain signals that something is terribly wrong with the body, depressive or anxious moods also signal that something is wrong or perceived to be wrong.[17]

DARWINIAN PSYCHIATRISTS SUGGEST that our emotions—good, bad, or off the charts—may not be pathological but instead important keys to our very success. Our psychological responses, therefore, may not be inappropriate, extreme, or disease states, but instead the reasonable reactions of a human mind that evolved under certain selective pressures and is faced with similar pressures today.[18]

In support of this provocative view, the Darwinian psychiatrists point to the fact that mental disorders have all the marks of universal human features that must have developed in accordance with natural selection. First, the most common mental disorders—such as schizophrenia, bipolar disorder, depression, and anxiety—are universal features of our species. Second, these conditions develop in a similar pattern among individuals.[19] For example, no matter the culture, schizophrenia most often first appears in young adulthood, while depression usually comes much later. Third, mental disorders interfere with social relationships, especially sexual relationships, which means the individual is not likely to pass along genes.[20] If mental illnesses were congenital diseases or viruses that had the same effect in our species, no one would hesitate to assume that evolution is at the root of their appearance.

Even more interesting, Darwinian psychiatrists believe that integrating the evolutionary view into psychiatry has the potential to change fundamentally traditional clinical practice. Nesse says that incorporating evolutionary thinking into talking with patients has led him to listen differently, and respond differently, and understand people more deeply. When I ask him point-blank how the evolutionary view might work when faced with, say, a depressed patient in the clinic, he answers quickly, "It depends on understanding what low mood in general is for. Unless you know what these defenses are for, and how they are regulated, you have no scientific grounding for making proper decisions about treatment."

By taking the evolutionary view, Darwinian psychiatrists are trying to change the very way that we, in Western culture, think about human emotions and mood. It may be that we are completely wrong in categorizing mental disorders as disorders at all. Perhaps periods of anxiety, sadness, panic, and sheer craziness are just part of the mental continuum that nature has provided to help us stay alive, make babies, and pass on genes.

If that's true, what possible value is there in schizophrenia, bipolar disorder, anxiety, or depression?

OUR SHARED REALITY

Humans are conscious beings, or so we think. No one can really explain what consciousness is, so the best we can do is offer the cryptic phrase, "We know that we know." At the moment, it looks like we are the only species with the kind of consciousness that allows us to take in the present, think about the future, and eat a sandwich at the same time. Chimpanzees are surely conscious of their environment and their group members, but no one knows what they think about the past, the future, or their world.[21] In that sense, humans seem to represent one end of a continuum of animal mental alertness. One thing we get for having big brains is the ability to check in with each other with language, and we have discovered that all around the world the vast majority of us, at the deepest and broadest level, think in similar ways. In other words, we presume a common reality.

Humans are so married to the idea of a common reality that it always comes as a shock to be around someone who seems to be on another planet. Years ago, while driving around the California town of Davis with a psychologist friend, we spotted a nice-looking young woman going for a walk. "See that girl?" he said to me. "Someone just turned on a radio in her head." He had treated this young women who was as shocked as anyone that there were foreign voices invading her mind.

Sometimes, without insider information, it's hard to tell that someone has crossed into another mental reality. In my teaching job at Cornell University, I am responsible for taking care of all the anthropology majors, and one day a young woman came in to complain about her advisor. She began with the usual complaints about the advisor not being available, and then moved into some personal problems she had been having during the semester. After about 10 minutes, I began to realize that she was completely fabricating some scenes, and that she and I did not share the same reality. What struck me about this situation later was the initial unconscious presumption on my part that we had a shared understanding, a presumption that I bring to all conversations with fellow humans, expecting that we will at least be somewhere on the same page. But I was terribly wrong in this case.

And when someone is not anywhere on the same page, they are considered the most mentally ill of all.

EVER SINCE FREUD, MENTAL DISORDERS have been conveniently divided into neuroses—such as panic disorder, depression, and anxiety—which are considered less serious because those afflicted may be unhappy but they have not slipped completely into another world, and the more severe psychoses, such as schizophrenia.[22] Schizophrenia is actually a catchall term covering a wide variety of symptoms, such as paranoia, delusions, hallucinations, and catatonia, which first seem to appear in adolescence or one's early twenties.[23] Schizophrenia of some variety appears in about 1 percent of any population; that is, it is universal. These are the people, no matter the culture, who seem to no longer share the common reality that the rest of us take for granted. By any measure, being schizophrenic is debilitating because the condition makes it difficult to function.

At first glance, it seems that such a destructive condition must be a pathology. Schizophrenia is obviously a mind gone wrong, and since the brain is a complicated biological organ, perhaps the mind has become ill in the same way that a heart, or liver, or lung can go bad.

But how does this "mistake" happen? It may be that schizophrenia is a recurrent mutation that periodically pops up in the human genome. In fact, research has shown that schizophrenia is more influenced by genes than most other mental problems, but that isn't saying much. Recent research shows that many genes probably affect the risk of various manifestations of schizophrenia, although there does seem to be a familial predisposition.[24] If one identical twin has been diagnosed as schizophrenic, the other twin is indeed at higher risk for the condition than if they were non-twin siblings, but that doesn't mean that both twins are destined to develop the condition.[25] Studies of twins suggest that only 60 percent of the variance in the appearance of schizophrenia can be explained by genes, which leaves 40 percent to environment, experience, and chance.[26] In other words, the presence of genes that put a person at risk for developing schizophrenia does not mean that person will necessarily develop the condition.

Given the universality of schizophrenia, and the relatively strong role that genes seem to play in its expression, this particular mental condition is appropriate for evolutionary speculation.[27] Is schizophrenia, with its varied expression, just a recurrent anomaly of the system, or is there some reason that this particular suite of symptoms appears over and over? Why does the human mind revert to a path that is so disconnected from reality?

Psychiatrists have always speculated that there might be some hidden advantage to being schizophrenic. The prevalence of the disorder alone suggests that some sort of genetic component of the condition might be favored in the gene pool. How this might happen is, of course, harder to figure out.

One group of researchers has suggested that schizophrenia might make a person more physically able to survive by being more resistant to stress, or viruses, or injury. Another camp has focused on possible social advantages.[28] Perhaps being schizophrenic means you don't need social attention and can get by within a family that has very little to give, or perhaps a psychotic episode means disconnecting from social pressures entirely, which might be advantageous at times in our highly social species.[29] Or maybe those with hyperquick minds were selected over evolutionary time because they could, when young, deal expertly with the social machinations of a highly social species. Later, in adulthood, these extrasensitive minds fizzle into psychosis.[30]

Evolutionary psychiatrists Michael McGuire and Antonio Troisi form another theoretical camp. They believe that most evolutionary explanations overlook the destructive effects of schizophrenia and that we should look in a different evolutionary direction.[31] Their rather novel approach is to focus on the fact that schizophrenia is characterized by a myriad of symptoms, including lack of personal insight, no empathy, and auditory hallucinations. All these symptoms, McGuire and Troisi point out, are similar: They are problems of information processing. For example, schizophrenia is marked by slow responses to stimuli, short attention spans, an inability to shift attention, and confused social interactions in which they simply can't participate. Schizophrenics are also unable to filter out useless information in their environment (such as the ticking of a clock or the talk of people in the hall). In schizophrenics, connections in the brain do not function as they should, and so input is misunderstood and life is a sensory jumble. McGuire and Troisi suggest that schizophrenics have low levels of cognitive ability and that they don't have one well-integrated automatic information processing system but several that bleed into each other. To compensate, schizophrenics withdraw into themselves, creating a world that can be controlled even if it isn't a happy place.

The key to an evolutionary approach to schizophrenia is the fact that the condition usually develops during adolescence, when young people are "finding" themselves and, most importantly from an evolutionary standpoint, looking for mates. For individuals with low cognitive skills and the inability to make clear sense of information about people and relationships, the world is an especially frightening place.

"It is easy to understand how individuals would be inclined to try to uncouple from a world that they experience as unpredictable and in which they consistently fail to achieve their biological goals," write McGuire and Troisi.[32] Unlike other mental disorders where sufferers cry out for help in many ways, schizophrenics move in the other direction—away from society and away from help.

DARWINIAN PSYCHIATRY OFFERS A NEW, perhaps more sympathetic, view of schizophrenia—as not just a disease, not just a disorder, not just a slippage of the mind into madness, but as a reasonable strategy given the genes and the situation. As Daniel Javitt and Joseph Coyle recently wrote, "Schizophrenia conspires to rob people of the very qualities they need to thrive in society: personality, social skills and wit."[33]

Since drugs are only sometimes helpful, and talk therapy is virtually useless with people who do not share our common reality, maybe sympathy is all we can offer, at least until we arrive at a better understanding of the causes of this debilitating disorder.

THE ANXIOUS SPECIES

While schizophrenia is deemed the most severe of human mental disorders, anxiety is the most common human mental affliction. According to the last two major surveys of mental disorders in the United States, 18 percent of the people in this country suffer from bouts of anxiety in any year. Worldwide, about 8 percent of the people treated in primary heath care report some form of anxiety beyond simple worrying.[34] People suffer from panic attacks that make them incapable of leaving home; some fall prey to compulsions, such as anxiously washing their hands over and over; others are brought to their psychological knees by a constant level of anxiety that permeates their lives and makes them afraid all the time. We are, it seems, a species of worriers.

Psychiatrists have identified various subtypes of anxiety, and they believe that each subtype has a different etiology and should be treated differently. Panic attacks, they claim, are not the same as agoraphobia, which is not the same as obsessive-compulsive behavior, which is not the same as social phobia, although they are all rooted in anxiety. The now traditional Western psychiatric response to anxiety is medication, which calms the mind and decreases the feeling of fear and worry. In other words, human anxiety in Western culture is now treated as a disease or disorder, a malfunction of the emotional system that can be managed. Yet anxiety in all its forms is so ubiquitous that it merits a second look from the evolutionary perspective.

Unlike traditional psychiatrists, Darwinian psychiatrists believe that anxiety may be an adaptive psychological defense mechanism that often serves people well, even today.[35] The idea of anxiety as a defense mechanism is based on the fact that humans are animals, and we share with other animals many of our basic reactions to stimuli.

In 1929 W. B. Cannon pointed out that a common suite of changes occur when any animal is scared. Cannon called this reaction the fight-or-flight response, in which heart rate increases sharply, glucose is released for quick energy, blood supply moves swiftly to the internal organs, sweat breaks out all over the body, and the animal hyperventilates, which raises the turnover in oxygen. All the senses are alive and rapidly taking in information and sizing up a situation to determine whether to run or lash out or act submissively.[36] These physiological changes in the face of danger are immediate and uncontrollable, and they make critical evolutionary sense. The body is on hyperalert status, making swift calculations about the force and direction of a threat; we might stand there frozen, sweating and hoping for the best, or run for cover. "An animal incapable of fear is a dead animal," write psychiatrists Anthony Stevens and John Price.[37] Without the basic instinct of fear, we'd be long dead as individuals and as a species; our ancestors would have been some other animal's dinner.

Anxiety, evolutionary psychiatrists suggest, is simply an extended form of the flight response and is related to the automatic response to a threat that in the past served our ancestors well. The evolutionary psychiatrists also point out that Cannon's detailed description of the physiological changes in the fight-or-flight response is the same for those that psychiatrists now use to identify a panic attack.[38] The heart races, palms sweat, the victim becomes paralyzed with apprehension—the course of a panic attack is very consistent and recognizable.[39] Human panic attacks today are, of course, not responses to man-eating lions or the threat of a boulder falling on your head. Instead, the human fight-or-flight response has apparently shifted from a clearly reasonable and necessary response to include reactions to psychological situations that may not look dangerous to anyone but the person who feels threatened. More significant, Darwinian psychiatrists also suspect that any anxiety that appears to be about nothing can be traced to a very real threat to survival or reproductive success.

Take, for example, the rich, anxious woman in Nesse's clinic. She is a very good example of what most people would see as self-perpetuated anxiety. Surely, she has everything and nothing to worry about. Her brain chemistry must be out of whack for some reason. Some psychiatrists would diagnose her as having an anxiety disorder and send her on her way with a prescription. Nesse thinks entirely differently. "Of course she is anxious," he says. "Her whole way of life is threatened. She sees these men divorcing their wives, taking up with younger women. Her marriage may be fine, but the threat is very real."

Apparently, we have only our large brain to blame for this vulnerability. Along with the great gift of consciousness and self-awareness that we gained with large brain size, we also received an intense awareness of all sorts of situations that are threatening to survival and reproductive success. The human mind can cast a very wide net into its past, present, and future and catch a million very real reasons to be fearful. A lizard sitting on a highway is not anxious because it has no idea that the car bearing down on it might cause its death. A person sitting on a road knows only too well the danger of cars, or the possibility of a hurricane, or the likelihood that someone might jump out of the bushes with a gun.

Humans are also very vulnerable to anxiety about social situations; attachment to others is so important to functioning in human families and groups that many people are often crippled by what is now called social anxiety.[40] Being anxious about going to a party, meeting new people, speaking in front of a crowd, or just going out to run errands makes sense if the weight of social interaction for individual survival is taken into account.

Nesse believes that our panic button is not necessarily something we should be ashamed of. He describes it as an inexpensive, but touchy smoke detector. A panic attack consumes only about 200 calories when it goes off, Nesse says, but is worth its weight in gold when the threat is real. Anxiety, Nesse therefore claims, is not a disorder; it's a psychological defense mechanism. "The anxiety disorders are not gross malfunctions akin to epilepsy," he writes. "They are only overly responsive defenses, much more like a tendency to vomit easily."[41]

According to this view, we are by nature anxious and fearful beings because we have knowledge that other animals do not have. We have the same reactive physiology, however, as other animals and a heightened mental awareness that causes very real fear and panic when threatened. Our alarm system works well, but who wants to live in constant fear? Anxiety is an emotion and like any emotion it can be triggered, exaggerated, or absent, and the key to a balanced mental life is in modulating that emotion.[42] "Anxiety is beneficial only if carefully regulated," writes Isaac Marks, a specialist in anxiety disorders who also considers the evolutionary view crucial to mental health treatment. "Too much disables. Too little anxiety leads to behavior that makes us more likely to fall off a cliff, be attacked by a wild beast, hurt by other humans, or to act in ways that lead to social exclusion."[43] Marks and Nesse also make the point that a system that evolved to help us can become a mental maelstrom within our large and active brain; we worry more about plane crashes than car accidents, more about the idea of dying from avian flu than from heart disease.[44]

All this theorizing about the evolutionary basis for human anxiety seems useless for the individual crippled by fear. How, in very real terms, does this approach inform psychiatry? The anxious woman in Nesse's clinic is a case in point. Her threat may be real from an evolutionary view, and her anxiety may make sense, but given that information, what can she really do except take medication or simply ride it out? Nesse, however, actually has a strategy to recommend: "In cases like this I suggest they go to college and they go to work. I tell them, your husband might stay and he might not, but you will have a life, and you won't have put everything in one basket."

How different it must have been for that woman to hear something so practical, so sensible, so life changing, from her psychiatrist.

"And you know what happens when she does this?" Nesse asks, relying on his observations of many similar cases. "She gets better."

DEPRESSION

We are each unhappy sometime in our life. Lost job, lost love, health issues, death of a friend or family member—life for everyone is a bit of a roller coaster. But for many, unhappiness is not just a temporary downer; it is a dark pit from which they seem unable to escape. More troubling, as mentioned in Chapter 1, depression as a diagnosed mental disorder seems to be a growing health concern around the globe.[45] If, like schizophrenia and anxiety, depression is another ubiquitous feature of the human species, might there be some evolutionary gain to being down?

THE TYPICAL SIGNS OF DEPRESSION are unrelenting sadness, loss of sleep, weight gain or loss, inability to see joy in anything, lack of initiative, a bleak outlook on life, and lack of self-esteem. Depression occurs twice as often in women than men, and hits women at a much younger age.[46] Although depression comes in many guises, and for many reasons, it is most often triggered by a life event—death of a loved one, a major life change, or a personal trauma—but sometimes it comes on its tiny bleak feet and walks all over someone's life for no apparent reason.

For most psychiatrists, depression is a pathological condition that is best treated with medication and talk therapy. Others take the view that depression is a symptom of an inability of the mind to regulate mood—up, down . . . and way down because the mind is out of control. Still others claim that a low mood in itself is not bad, and may even be useful, but when it continues for long periods or becomes morbid, something is indeed wrong.[47]

Darwinian psychiatrists are especially interested in depression. It's a negative mood state that appears in all countries, societies, and cultures, and the symptoms are familiar—everybody, simply everybody, recognizes depression. We all know what it means to be unhappy, or sad, or down, and so we can relate to the possibility of that dark hole. How could an emotion so bleak be of any use to anyone?

The evolutionary perspective has produced several conflicting hypotheses to account for depression. For example, researchers have suggested that depression might be a way for people to disconnect from society and conserve energy during a time of great stress.[48] Others believe that the greatest stressor would be a threat to an important social or personal attachment[49]—a wife leaves, or a child dies, or colleagues betray a person. Depression might be triggered because the human mind evolved to be attached, and the loss of a relationship represents a decrease in future survival and reproductive success. Edward Hagen believes that depression is a sort of manipulative strategy to get attention and resources from others.[50] Depressed people, Hagen thinks, cry out at every level for attention and attachment, and so depression might have evolved to bring the depressed person back into society. Psychiatrist John Price suggests that the social hierarchy that frames human interaction is really to blame.[51] According to Price, people are constantly involved with conflicting social situations, and they have the option to escalate and fight back or de-escalate their behavior and mood, feel humiliated, and become depressed.[52] In support of this notion, Paul Gilbert and colleague Steven Allan tested this idea by engaging 302 undergraduate students and 90 depressed patients to complete a variety of questionnaires about themselves. The researchers discovered that depression, and feelings of being trapped, were indeed associated with a sense of defeat in some real or imagined interaction.[53]

Price maintains that it's not the mood itself, but the flexibility inherent in our emotional system that was a great advantage to our ancestors, because it allowed them to settle rapidly interpersonal disputes. Today, Price explains, life is much more complex, and we are constantly faced with conflicts that are not so easily resolved. Some conflicts are with parental ghosts, some are with people who can't be reached for other reasons, and as a result de-escalation is rampant. For Price and his colleagues, depression is all about how a person sees his or her place within a network of relationships; for them, depression is an interpersonal issue and all about social competition. Some win and some lose, and when you lose, it's better to withdraw and be depressed than fight back and take the chance of dying, whether literally or metaphorically. Therefore, we have evolved a brain mechanism that lowers mood, but keeps us alive when the bad times come.

I ASK RANDY NESSE ABOUT depression late in the day. We've gone for a walk through the streets of Ann Arbor, bought some candy and Cokes to elevate our mood, and are back in his office where the phone rings and students pop in to drop off papers. We are both surprisingly cheery even though we have spent most of the day talking about mental illness. I've been expecting our talk to get difficult because talking about depression is, well, depressing. After all, I have family and friends who have struggled with depression for years. Some are clinically depressed, meaning their mood is extremely low and intractable, while others appear to float continually below happiness and right on top of worry. And nothing—not medication, not therapy, not even time—seems to do much to make these people feel better for very long.

Surprisingly, Randy Nesse hands me a perspective that is more optimistic than I could have imagined. He says that we have been shaped by natural selection to pursue certain goals because moving forward is useful for our survival and reproductive success. The problem is that there is now a mismatch between how our minds evolved to pursue goals and the reality of achieving those goals in our modern context.

"What influences mood is the gap between what people have and what they want," he explains. "We have the capacity to foresee the future and pursue large goals. What actually creates lower or higher mood is a person's rate of progress toward those goals. If you anticipate that you are on your way to a valued goal, you're fine, even if you are cold and poor and shivering."

"We are designed by evolution to pursue goals that take one or two days, maybe three weeks. Our ancestors would go three days looking for nuts and when they didn't find any, they would camp for a few days and not waste their time looking for nuts. It's a very sensible thing. Now we are pursuing goals that take one month, two years, 20 years."

As I think about the role of goals in my own life, Nesse reaches into his example bag and throws out a series of bad possibilities aimed at my personal universe: "What if you submit a paper to a journal four times and it never gets accepted? What if your long-term marriage is bad? There are no exits, no alternatives? What we get is feeling hopeless, worthless, pessimistic, no energy. People have these goals and they can't give them up, and it's just as abnormal as dehydration from diarrhea."

I wonder whether Nesse isn't too focused on the Western ideals of being successful, making piles of money, and buying lots of stuff. But he squelches those thoughts by taking the idea of hope a step further. "We really are shaped by our genes to pursue these goals for their own benefit," he continues. "I think it's probably an elaboration of the same mechanism that gets a bumblebee to move from flower to flower, and the same mechanism that gets a bumblebee to go home at the end of the day, and the same mechanism that gets a bumblebee to hide in its burrow during winter. There are all these different levels of regulating motivation, moment to moment, day to day, season to season, and I suspect that we'll eventually find that the very same or related brain mechanisms are regulating motivation in humans."

Depression for Nesse, then, is the loss of hope.[54] "Up" mood is having hope, depression is loss of hope, and the trick is to get someone out of a depression, no matter the circumstance, and restore that person's hope.

Nesse also adds a Pollyanna observation that puts a happy evolutionary spin on unhappiness. "You don't get depression unless you have underlying hope," he says, hoping to give me some hope. "People aren't hopeless unless they are hopeless about something. There is always some goal that you haven't yet given up on if you are hopeless about it," he says, smiling. Ever the evolutionist, he also adds a Darwinian tag line: "I think natural selection shaped our mental mechanism to keep us from jumping ship too soon, even when things are not working." In other words, we are designed by evolution to hope for something for a long time, and then to feel hopeless, because hope is integral to our survival.

Full of hope, surprisingly chipper for someone who thinks about mood disorders all day, Nesse also thinks there could be real changes in the way Western culture treats mild and chronic low mood. "An evolutionary approach in psychiatry could lead us toward a better understanding of the client's motivation," he says. To that end, he wants psychiatrists to ask their clients a whole new range of questions. "How big are the goals people are pursuing?" he suggests. "How long do they need to be pursued? And if it doesn't work out, do they have alternatives? We need to ask not just what caused the current depression but also what a client has in terms of family, friends, social resources."

In the clinic where he sees psychiatric patients, Nesse now asks those sorts of personal questions. Beyond "How do you feel?" and "When did you start feeling depressed?" he now adds, "Are you pursuing some life goal that you can't reach and can't give up? Are you afraid of losing something that is essential to your life?" In asking those questions, he hopes to bring hope to people down in the black hole of despair.

EVOLUTIONARY PSYCHIATRY IS NOT the answer to eradicating mental illness. But it is, I believe, a piece of the puzzle to understanding why our species is so riddled with anxiety, unhappiness, and insanity. If what the Darwinian psychiatrists suggest is true—that mental disorders are not necessarily pathologies but strategies, adaptations, or responses that make evolutionary sense—then we have a whole new way of looking at human unhappiness, which means we have a whole new way to help.

It also means that suggesting that mental illnesses are primarily biological, as the Western medical model does, is much more complicated than merely bad brain chemistry. It means there are all kinds of ways of producing that brain chemistry, all sorts of influences that might contribute to a person's outlook on life.

Those influences may have been thousands of years in the making, or they may have happened last week.

CHAPTER 3

THE MINDS OF MONKEYS

I'M STANDING IN AN ANIMAL RESEARCH LAB IN POOLESVILLE, MARYLAND, dressed entirely in blue paper—paper shower cap, paper shirt and pants, even blue paper shoes. My face is covered with a plastic shield, and my hands are gloved. This getup is required because I am holding a newborn rhesus monkey, a tiny ball of fur with brown button eyes and hair that sticks straight off the top of its head.

The baby monkey wiggles around in my hands, grabs my thumbs, and makes "coo-coo" noises that are heartbreakingly endearing. I look into his eyes and think that this little guy might just be the cutest thing I've ever seen. He is also a research animal in an ongoing scientific study. No one will give him drugs, or try out some kind of new surgery technique, or ever physically harm him in any way. Instead, as he grows this rhesus monkey will be put into various social situations and simply watched—very closely. He is an animal model for human mental health.

The idea that monkeys could be stand-ins for human behavior and human ills is not new. Monkeys and apes, our closest genetic relatives, have been used for a long time in research on human diseases and medical advances. And for decades, nonhuman primates have been observed in captivity and in the wild to understand the basis of human behavior.[1] Jane Goodall, for example, went to Tanzania in the 1960s explicitly to observe chimpanzees and compare their behavior with that of people.[2] "If you see a behavior in chimpanzees and you see it in humans, you can be assured that the behavior comes from a common ancestor," Goodall has said in lectures. In that statement, Goodall is making an evolutionary connection between what chimps do and what humans do. She believes that both species evolved from the same stock and that human behavior is not unique, because we are the same as chimpanzees in many ways.

This is not to say that everything about a nonhuman primate is relevant to human behavior.[3] Indeed, other primate species (lemurs, monkeys, apes) have evolved along different paths for different reasons. But psychologists and anthropologists believe that monkeys and apes have a lot to tell us about the roots of human behavior. By observing these animals and comparing their behavior with that of humans, we can elucidate what is common in all of us primates and what, if anything, is different or special about humans.

Because nonhuman primates share our genetic history, they are also considered models for the evolution of the human mind, of our emotions and mood states. Humans may have bigger and more complex brains, but we can also see ourselves in a baby monkey as it coos for its mother or a chimpanzee as it hugs and kisses a friend. In that sense, the way a monkey or ape reacts to an emotional experience has implications for how humans feel. Because of their genetic relationship to us, and the fact that they are so similar socially and emotionally to us, nonhuman primate species represent a potential psychological window into the deepest recesses of the human mind.

THE MONKEY MODEL OF UNHAPPINESS

In the 1950s a psychologist named Harry Harlow of the University of Wisconsin was running a primate research laboratory and trying to breed enough rhesus macaques to keep his various projects going.[4] Following the fashion for human births in hospitals at the time, the lab protocol for breeding monkeys was to take the monkey infants away from their mothers soon after birth and rear them by hand. The idea was to make the infants' environment safe, clean, and scientifically controlled. Soon everyone at the lab realized this protocol wasn't working at all—the hand-raised monkeys were traumatized by the separation from their mothers and the lack of interaction with other monkeys. More shocking, the researchers noticed that when the early-separation monkeys reached adulthood in a few years, they were unable to have sex and unable to parent.

Harlow and his students realized they had stumbled across a psychological experiment, a behavioral protocol that seemed to change the very moods, emotions, and feelings of monkeys, and that deeply affected their psychological development.

Harlow's findings made him a rebel. In the late 1940s America was in the grips of the Behaviorism School of psychology, which advised parents to keep children at arm's length if they wanted their kids to grow up "normal," that is, self-reliant and independent. At the same time, psychiatry was wallowing in psychoanalysis, focusing on the unconscious and sex, and there, too, parents were scared of screwing up their children with inappropriate affection. The focus was on the individual as a separate, independent unit, not on the person as a part of a larger social network.

Harlow suspected that psychologists and psychiatrists were completely wrong. If emotional independence was natural for primates, including humans, why were his nursery-raised monkeys so depressed and anxious?

Harlow's initial insight was underscored at the time by the writings of psychiatrist John Bowlby.[5] Bowlby was trained as a Freudian analyst, but he broke from the fold and was convinced that humans, like rhesus monkeys, very much needed strong early connections with nurturing and responsive caretakers in order to grow up emotionally healthy and that unconscious sexual thoughts had nothing to do with it. After reading Bowlby's work, Harlow decided to use the rhesus monkey as a psychological animal model to prove that both monkeys and people need close attachments in order to be sane.

In Harlow's initial experiments, infant monkeys were removed from their mothers and put in a cage with either a wire form of surrogate mother, which held a bottle, or a cloth form that provided nothing but a soft place to cling. The researchers found that the infant monkeys overwhelmingly preferred the cloth mother: They clung to it all day, crawling over to the wire mother only when they were hungry. It was clear that given a choice, a warm, soft touch, not nourishment, was the most valued connection for the little primates. These early experiments demonstrated that young primates—and presumably that means young human primates as well—are desperate for interpersonal attachment and nurturing love, from at least one primary caretaker.

For the next several decades, Harlow and various students and colleagues designed endless permutations of the rhesus infant separation protocol.[6] Babies were taken off their mothers right after birth and placed in a cage alone. These monkeys exhibited behaviors that were disturbing. They became hostile and detached and displayed repeated anxious movements like a person in solitary confinement. They often huddled in a corner in despair.[7] The researchers also discovered that the same kind of unhappiness could be elicited in an adult monkey by taking it out of its family group and putting it in isolation for days on end.[8]

Harlow maintained that his depressed, isolated monkeys were suffering from a kind of learning deficit—they simply didn't know how to interact as monkeys. A lack of understanding of the social world made them, and presumably could make any primate, depressed misfits.

Harlow and his student Stephen Suomi then decided to push even further the possibility of a nonhuman primate model for depression by taking isolated infant monkeys and putting them into a metal upside-down pyramid with no way for the monkey to see, hear, or feel anyone at all.[9] These monkeys very quickly went crazy. They chewed on themselves, huddled in apparent psychological agony, and became immobile. Harlow called the device the "well of despair," and one look at the photographs of little monkeys crouched in the fetal position at the bottom of the cone confirms Harlow's label. The monkeys were, by anybody's definition, severely depressed.

Harlow set out to make a monkey model for depression, and he did.[10] The idea of having these animals stand in for human mental illness might be unsettling, even shocking, but still the experiments were very much a success in terms of scientific value. Isolating an infant monkey destroyed its emotional and psychological development, and isolating an adult monkey made it insane. "The findings of the various total-isolation and semi-isolation studies of these monkeys suggest that sufficiently severe and enduring early isolation reduces these animals to a social-emotional level in which the primate social responsiveness is fear," Harlow wrote in 1965.[11] Fear took over because these monkeys had no social life, no attachments, no family, no friends. Being isolated from others turned out to be the worst thing you could do to a monkey psychologically, and presumably it is the worst thing you can do to a person.

Harlow's isolation research was also about love.[12] Love, of course, can only occur with others.[13] These painful experiments demonstrated that we primates are profoundly susceptible to the loss of social attachments because we are designed by evolution to be connected to others of our species from the moment of birth; being antisocial goes against our very nature. Taking away those attachments results in a psychological death.

Harlow's other aim, besides producing a monkey model for depression, was to figure out ways to heal these animals. In doing so he hoped to give human clinicians some idea of how to treat mental illness. The researchers did the obvious—they reversed the process and began to introduce the possibility of social attachments to the isolated and depressed monkeys.

First, they placed peer therapists, that is, other young monkeys that were especially affectionate, into the cages with the disturbed subjects for a few hours at a time. Eventually, the depressed monkeys began to respond to the touching and hugging, and over time they started to act like any other monkey.[14] In addition, female monkeys that had been reared in isolation and later gave birth as adults seemed to be socialized by their own infants; the unrelenting contact by an infant seemed to bring out the best in these disturbed females.[15] The animals that had been in the "well of despair" were the hardest cases to treat, and it took psychologist Melinda Novak many months and a lot of patience and consistent, targeted therapy to bring those very disturbed monkeys back to a level where they could interact in a normal way. But even the most deeply depressed monkeys eventually responded.

The work of Harlow and his colleagues speaks to the therapeutic process in humans. Maybe the best mental health therapy is not individual talk therapy but sessions conducted in a more complex social milieu. These studies of monkeys also suggest that even in the most intractable cases of unhappiness there is hope.

Today, we accept as common sense the notion that babies should be with their mothers and that social attachments are imperative for developing and maintaining a healthy mental state. In fact, we are indebted to Harlow and his colleagues for such common sense. Before those experiments, psychologists and psychiatrists in Western culture believed that mothering made no difference and that infants didn't really need contact to be psychologically healthy. Even today, long after Harlow's initial experiments, many parenting experts and parents push for independence and self-reliance for infants and children.[16] Nonetheless, Harry Harlow's isolated baby monkeys stand as the most important animal model to demonstrate that depression can arise in primates through loss of an expected social attachment.

This work also points out that primates, including humans, need to exercise their social chops to be mentally functional. Harlow knew his monkeys were unhappy because they were fearful, anxious, and depressed when they were alone. Even limited periods of social deprivation left their mark, and the animals did not snap back immediately when presented with the opportunity to be social. There, too, is an important lesson for understanding the basics of human mental health.

A PRIMATE'S EMOTIONAL NATURE

It's hard these days to understand how revolutionary Harlow's experiments were. In hindsight, with over 60 years of research on primates in the lab and in their natural habitat, it seems obvious that socially isolating a primate baby would cause serious emotional harm. We now know that the most significant feature of our taxonomic order—the primates—is that we tend to live in groups characterized by intense interpersonal interaction. In contrast, most mammals are rather solitary, and they like it that way. But we primates thrive on being together. From the lemurs on Madagascar to the bonobos in the forests of central Africa, our primate cousins gather together and interact. Of course, many animals, such as antelopes or zebras, aggregate, too, but what makes primates different from those herds is the kind of interaction that characterizes the groups. A rhesus monkey, for example, is born to a female about four years old. Right away, the newborn rhesus will be flung into whatever social dance might be going on at the moment, with aunts and cousins pushing in to take a look. Throughout her life she will fight, groom, play with, and have sex with individuals that she knows well.[17]

Primates have been selected by evolution to have special abilities that allow them to manage their social connections.[18] For example, we are able to recognize each other and classify individuals into discrete, socially meaningful categories. We also mentally put others into "in" groups and "out" groups and decide which category is best. Similarly, rhesus monkeys prefer their mothers, cousins, aunts, uncles, and whoever else is part of their matriline, the female kin who stay in the group (the males typically emigrate to other groups). Macaques also favor kin over nonkin in fights and spend more time with relatives.

Even rank order, which is strict among many macaque species, can be traced through kinship lines.[19]

The same kind of nod to kin is found in all other social primates—attachment to family is clearly in our genes. Favoring kin makes evolutionary sense, of course; we share genes in common with relatives, and so when there is a choice, it should be for our kin.[20] But more significant for our mental health, the ties of kinship that evolved to help us stay alive have emotional and psychological meaning as well. We seem to need a social network that is broad, deep, and strong, and without those ties we may lose our bearings and actually go crazy.[21]

We also know from observing these animals for decades that all primates are blessed with a keen sense of social organization that goes beyond the bonds of kinship. Many primate species construct an overall troop social hierarchy that determines who gets to do what with whom. These hierarchies make for an organization that each animal recognizes; every member knows exactly where every troop mate fits into that organization, although individuals change position all the time.[22] Our kind is also good at using social knowledge to fit in with and manipulate others in daily life; being good social animals means knowing whom to support in a fight, how to sneak around dominant males, and when it's just the right moment for grooming.[23]

Our socially manipulative side is counterbalanced by the fact that humans also have evolved the capacity to express sympathy and empathy. Primatologist Frans de Waal has written extensively about the evolutionary roots of caring, and he makes a good case for accepting that chimpanzees, and other primates, have a rudimentary sense of empathy and sympathy.[24] According to de Waal, when a chimpanzee female makes peace between two angry males, or when troop mates hug each other in times of fear, the animals are showing the same kinds of emotions found in humans. Because evolution has opted for primates to gather in groups, we have used our big brains to feel not just for ourselves but for others as well. So along with nurturing love, kinship bonds, and social organization, empathy and sympathy are necessary for intimate group living.

The need for group living—and all its dynamic motion and complexity—is an aspect of who we are as evolved beings. We are social animals down to our core, and both the social context and our compelling need to be connected are a part of our nature that cannot be changed. It is too deeply embedded in our makeup.

THE NATURE AND NURTURE OF
MONKEY MENTAL HEALTH

Back at Poolesville, I reluctantly hand over the baby rhesus monkey, shed my blue paper costume, and cross the grassy compound to see where young experimental monkeys go when they graduate from the nursery. My guide on this part of the tour is psychologist Dee Higely, a tall man with a buzz cut and a warm smile who looks like the academic version of Gene Hackman on a good day. We chat about his work on risky, impulsive behavior and its role in alcoholism in monkeys, which he has been exploring for the past two decades. Higely then opens the door of a nondescript building and I am greeted by the sight, sound, and smell of a huge gang of juvenile rhesus monkeys. There must be over 50 of them in this big, noisy room.

Many years ago I worked at the University of California, Davis, Primate Center, so the antics of the two-year-old rhesus monkeys in this hall before me are familiar. Like any young kids, they are a hoot. As we walk to the large caged enclosure, dozens of them scramble right to the front, curious and excited because someone new is here. They stare at me, poke their hands and feet through the wire fencing, and call out in both fear and joy. I know from experience that they'd love to get a hold of my hair and my pen, or use me as a springboard. The whole group is in constant motion—they approach and recede, touch and run away, leap and screech—and Higely and I know that every single movement has social significance to these animals, just as it does in human groups.

This room full of juveniles is yet another experimental living condition used at Poolesville. Some of the monkeys will eventually move to outdoor spaces, where they will live with adult monkeys. Others will be shifted to cages with smaller groups. These different social groupings are there for good reason—mostly to control and study the social experiences of the animals—but none of the arrangements are true rhesus monkey groups. A naturally formed rhesus troop is composed of many adult females, a few males, and many kids. Higely tells me that besides the monkeys here at Poolesville, he and his colleagues are working with free-ranging rhesus monkey groups in a place called Cayo Santiago, an island off the coast of Puerto Rico, where Indian rhesus monkeys were imported in the 1940s, and another group on Morgan Island off the coast of North Carolina. Obviously, these troops are not free ranging in the sense that they might be in their native habitat of India or Pakistan, but both colonies still allow the researchers to test their theories on naturally raised animals who are in naturally formed rhesus monkey social groups.

Using the socially manipulated babies raised at Poolesville and the more free-ranging groups, an extensive network of psychologists, psychiatrists, biochemists, and geneticists are collaborating to move far beyond what Harlow started. In the past few years, in fact, pieces of the puzzle that Harry Harlow was beginning to identify and put together 60 years ago are now falling into place to reveal a vivid picture of primate mental mood and behavior.

PSYCHOLOGIST STEPHEN SUOMI HEADS the Comparative Behavioral Genetics Section at Poolesville. Suomi worked with Harlow years ago at the University of Wisconsin Primate Lab, and he knows all too well how to make an unhappy monkey. What interests Suomi now is how the path to unhappiness occurs within the context of an individual monkey's life, and the interaction of genes and experience. I've known Suomi and his work for years, so after leaving Poolesville I call him, and we spend some time catching up on what he's doing these days. Apparently, Suomi's research is moving farther, faster, and in more directions than he ever expected. "Things are really different these days," he tells me, with excitement in his voice.

Suomi starts with the basics of what it means to be a social animal. Social challenges, he explains, are part of life; the average day is full of tense interpersonal interactions and problems to be solved, and some days are worse than others. People and monkeys respond to these challenges in different ways. Some individuals are able to cope quite well, rolling with the punches, moving through the day without too much trouble. Others have a harder time. What makes life easier for some and a constant struggle for others?

"Some components of these reactions are heritable," Suomi says. Given the current fashion of linking human mood and behavior to genetics and biology, this statement is not a surprise, but I didn't expect to hear that the details of how individuals react have been worked out using rhesus monkeys.

Negative mood and behavior, according to Suomi and others, come in two flavors. "Across the board, about 15 to 20 percent of rhesus monkeys in our population appear to be excessively fearful and anxious," he tells me. "We call them uptight monkeys." Interestingly enough, the rate of anxiety among human adults is about 15 percent of the population as well. The anxious monkeys also look very much like the subgroup of children studied by Harvard psychologist Jerome Kagan. Suomi and Kagan are convinced that they are looking at the same kind of individuals: kids who react to situations rather than respond, and who are scared even when a challenge is mild.

The fearful and anxious monkeys have a physiological reactive profile that indicates they are under great duress in situations that wouldn't phase a better-adjusted monkey. With the slightest provocation, their hearts race, the stress hormone cortisol floods their bodies, and there is a rapid turnover in the neurotransmitter norepinephrine. A fearful and anxious monkey is almost perpetually posed to fight or run. Since the researchers at Poolesville have the genealogies of all their baby monkeys, they have discovered that this behavior can be tracked over generations.

At the same time, however, genes are not everything. These researchers know from observation that even though a mother is anxious and fearful, her sons and daughters may turn out much mellower, and the converse is true as well. "If these genetically anxious monkeys have very good nurturing mothers, they are basically buffered from anxiety and fear," Suomi explains.

That's one subgroup of mental type. Another 5 to 10 percent of monkeys in the lab and in the free-ranging groups—and about the same percentage of people—are impulsive and aggressive. Again, being impulsive and aggressive seems to have a strong genetic component, but the experimental monkey work demonstrates that this sort of behavioral profile can be modified substantially by early experience. "Bad mothering makes these individuals much worse," says Suomi. "But if they have good mothers, it doesn't make any difference what genes they have; they're protected."

In other words, monkeys that grow up with mothers who love them, bond to them, and teach them the social rules of being a normal monkey are fine, even if they have inherited genes that predispose them to be anxious, fearful, impulsive, or reactive. But if they have been taken from their mothers, or if they had an ill-suited, rejecting mother, these subjects will sink into monkey unhappiness, fighting with others and always in harm's way.

The news that monkey behavior that might be determined by genes—especially genes that can influence negative moods and behaviors—but also modified by early experience has been revealing. Even more exciting, scientists have begun to discover exactly which genes put the monkeys, and people, at risk for unhappiness.

RESEARCHERS HAVE KNOWN FOR A long time that the amount of a neurotransmitter called serotonin circulating in a person's brain has something to do with mood.[25] People with lots of serotonin seem to be happier than those with depleted supplies; many people on antidepressants, which elevate levels of serotonin, claim their mood improves. No one is quite sure why serotonin—rather than any of the 141 other identified neurotransmitters—is important, or why more of it elevates mood.

In 1996 a group of German scientists reported that they had found an association among three factors: (1) the presence of a variation in a region of chromosome number 17, (2) the amount of serotonin available to the brain, and (3) anxiety-related traits. The site of the variation is not one that codes for making serotonin per se, but it regulates how much of the neurotransmitter is allowed to bathe the brain. The serotonin transporter gene comes in two forms, the "l" (long) version and the "s" (short) version—these labels actually refer to the length of the DNA of each form. Everyone has two chromosomes number 17, which means you can have two short versions of the gene (s/s), one short and one long (s/l), or two long versions (l/l).

In a sample of 505 people, the researchers discovered that having just one "s" form of the gene, which also meant the subjects had less serotonin, accounted for much of the appearance of anxiety-related personality traits, such as being neurotic, tense, suspicious, and worried. Presumably there is a connection between the serotonin transporter genes and anxiety.[26]

A flurry of studies on human groups followed, but the results were inconsistent. Sometimes there was an association between the short form of the gene and anxiety and sometimes not. No one could tell why the results were coming in so mixed.

Under the best of circumstances, it's difficult to study people's psychological moods, especially if you want to work with big, convincing datasets. It's also difficult to actually track people's moods. When self-reporting mood, people are notoriously fuzzy. Even psychological testing of humans can be a shaky way to get at someone's overall psychological state because emotional states are dynamic and a one-shot test can't always track mood changes. But at Poolesville there was an animal model with both long-term genetic records and objective reports of behavior. So in 2002 Alison Bennett of the

Poolesville group looked at the serotonin transporter site in monkeys and found that they, too, had short and long variants of the gene, and in the same frequency as the numbers in human groups. This finding was sort of serendipitous. We might expect chimpanzees, more closely related primates, to have the various forms of this gene, but they don't. Only rhesus monkeys and the South American capuchin monkey seem to have the same rates as humans of the long and short forms of this gene.

Bennett then looked at 132 rhesus monkeys—some had been taken away from their mothers at birth and had lived a life of social stress, and others had had a relatively normal monkey life. Some in each group carried the short version of the transporter gene. Bennett looked at the amount of circulating serotonin for all the animals and discovered that for those with good mothering it didn't matter if they carried one short version of the transporter gene. Everybody in that group had about the same amount of serotonin in their spinal fluid. But for those monkeys that had been isolated in the nursery and then raised with peers under social stress, genes really did matter: Carrying even one version of the short version of the gene meant significantly lower serotonin in the brain.[27] In other words, individuals who have just one short version of this gene are at risk for developing both depressive and anxious behavior, but only if they also have poor early experience. Otherwise, they are fine.

In a moment of scientific confluence, another study, this one on humans, showed that indeed the determining factor wasn't just the presence of this or that gene, but the impact of stressful events over a life course. Like the monkeys, people who have one short version of the transporter gene just don't do as well with stressful events. In that sense, they are unequipped to deal with the loss of a loved one or some sort of accident, and if they are hit with recurring bad luck, depression ensues. The very same people with one short allele might get through life without being touched by something awful and without ever knowing they were at risk for depression.[28]

The Poolesville researchers rapidly made the connection between the short version of the transporter gene and other negative behaviors in rhesus monkeys, such as drinking too much alcohol, or an excessive stress reaction to separation from their mothers.[29] In all these cases, there was an interaction between the short form of the serotonin transporter gene, upbringing, and negative mood or behavior.

It would seem that having "bad" genes would be devastating for both monkey and human populations, but the "s" version is ubiquitous. For example, half the Caucasian population carries a short gene.[30] Why do these genes that risk causing negative mood and behavior appear at all? Do the monkeys have an answer for that, too?

According to Suomi, understanding how rhesus monkeys make it in life probably holds clues to why the short form of this gene has been maintained over generations in humans. "Rhesus macaques are the most geographically widespread species of primate after humans," Suomi explains. "They live in every possible type of habitat and are the most numerous of all the monkeys." If you count heads, this is certainly true. In contrast, there are only four ape species, and they are restricted to very specialized habitats; their numbers could easily be counted on an abacus. Rhesus macaques, on the other hand, are practically everywhere, from Pakistan to China, and they can survive where no other monkey would be caught dead. In fact, researchers have named species that can invade an environmentally hostile place and do well "weed species," and Suomi thinks that both rhesus monkeys and humans are weed species that have flourished under difficult conditions.

Maybe the short transporter gene has aided our weediness. On the research islands studied by the Poolesville group, impulsive, aggressive males tend to be kicked out of their troops around three years of age, well before the normal age of emigration (about five years), and most of these early emigrants die within a year.[31] These are also the males with low serotonin levels.[32] Females who have this type of personality tend to be at the bottom of their hierarchy, living not very pleasant lives. These females survive and have infants, but they tend to be poor mothers, and they pass their poor mothering style along to their daughters. On these geographically restricted islands—monkey paradises where food is provided and there are no predators—the genes underlying aggressive, anxious behavior appear to be a liability. In the harsh reality of the wild, these very same animals might be the ones who push into new environments, battle with the elements or other monkey troops, and survive. So what are considered "bad" genes in one environment could be advantageous under other circumstances.

What about the other subtype of monkey, those that are fearful and depressed? Surely being afraid and unhappy would be a liability in a species known for its collective aggressive attitude. Wouldn't these monkeys be behaviorally eaten alive by their nasty troop mates? Suomi thinks not.

"Males with this behavioral profile actually emigrate late, postpone it until [ages] 7 or 8," he explains. "And because the best predictor of male survival during male emigration is size and weight, if you postpone emigration until you have finished the adolescent growth spurt, it's at an advantage for survival. For these guys, being fearful is advantageous during that period."

The same is true for fearful female rhesus monkeys. "Females who have this kind of profile and are in stable social groups are actually really good mothers, even with excessive fearfulness," Suomi continues. "If they are in unstable social groups, their maternal behavior falls apart, but it actually may be protective in stable social groups." Most significantly, Suomi points out, rhesus monkeys, like humans, can carry both the long and short forms of the serotonin transporter gene. Depending on what is going on socially, and where the troop might be living, there is clearly enough genetic variability among rhesus monkeys to produce a variety of personality types, and so as a species they are eminently adaptive.

Echoing the words of Randy Nesse in Chapter 2, what we think of as bad genes that result in low mood and inappropriate behavior might be just the ticket under certain social or environmental circumstances. "It's the capability to have that kind of behavior," Suomi says. "The genes bring one part, and then the environment provides a forum in which these things are going to be useful or devastating." For species such as rhesus macaques, or humans for that matter, our ability to inhabit all the nooks and crannies of the globe may be underwritten by genes that affect our mental outlook, motivation, mood, and social skills.

The significance of this research to the way we view mental illness is profound. The monkey data and the human research both point to the interaction of genes and experience. That's what excites Suomi the most.

"We're looking for specific gene/environment interactions and we're seeing them all over the place," he says. "We can pick these interactions up in the first month of life and see early neurobehavioral markers. We pick it up in terms of physiological markers, can pick it up in terms of alcohol consumption, and pick it up in terms of hormone activity." Bad genes translate into sad mood and inappropriate behavior only if a caring someone is not around to modulate social development. Experiences as an infant, child, and adult all play a role in the expression of whatever genes are there.

"This is a dynamic system," Suomi says. "The genotype is just the start of it. What is really important is the regulation of those genes, and what's doing the regulating. Environment, especially the social environment, one way or another, is regulating these genes, turning them on and off."

And because of these monkey models, it is no longer possible to say that depression, impulsivity, and anxiety are "genetic" or "biological" in some simple way. Individuals have particular genes that make them vulnerable or resistant, depending on what experiences they have had in life and what circumstances they face.

As AN ANTHROPOLOGIST, I WONDER mostly about the influence of the environment that intimately interacts with one's genes. For monkeys, and presumably humans, a nurturing maternal presence early in life is surely important. Social interactions are clearly as formative for monkey mental health as they are for people.

What else about one's environment might make a difference? For that answer I turn to some of the newest and boldest research on mental illness.

CHAPTER 4

THE HAPPY FAT

I AM SPEEDING THROUGH THE BACK ROADS OF MARYLAND IN A 1972 RED convertible Chrysler Le Baron. The top is down, the sun is shining, and the Rolling Stones' *Beast of Burden* blasts from the radio. It doesn't get any better than this.

My driver, psychiatrist and National Institutes of Health researcher Joseph Hibbeln, turns to me with a big grin. "How happy are you?" he yells over the music. "Could you *be* any happier?"

With his round cherubic face, a shock of brown hair that slides down his forehead, and that grin, Hibbeln seems to be the happiest man on Earth. Underlying that happy-go-lucky manner lurks a dedicated scientist concerned about the growing number of cases of debilitating depression around the globe. Even now, in the midst of a moment of shared glee, he can't help himself from bringing us back to the point of his research.

Hibbeln turns down the music, glances at me from the sides of his Ray-Bans, and says with as much gravity as one can have while breaking the speed limit, "You don't really need the convertible to be happy, of course. The sunny day helps. The Stones help. But that fish we had for lunch is what you really need." Although the idea of a fish lunch as the root of all happiness might seem incongruous while zipping along in a convertible, I know Hibbeln is completely serious.

Earlier in the day, in his office at the National Institutes of Health, Hibbeln had flipped through chart after chart showing the correlation between the consumption of seafood and rates of depression and other negative moods. For the past 15 years Joe Hibbeln and others have been working on the role of essential fatty acids in human health, and they are convinced that the typical Western diet is not just bad for our heart and blood pressure, it's also apparently bad for our brains. The issue is not that we eat too much fat; the big news is that we eat too much of the worst *kind* of fat. We wolf down processed foods that contain hydrogenated soy oil, munch on snacks fried in corn oil, and ingest too much margarine. In doing so, we take in an excess of a type of essential fatty acid called omega-6. At the same time, we push away foods like fish and olive oil that are full of another kind of essential fatty acid called omega-3. This imbalance of the two kinds of fats with the skew toward omega-6 is important because the human brain is designed to operate on a 50:50 balance of each; but our modern Western diet has filled our brains with an imbalance of fats that actually hinders the brain's electrical connections and lowers mood.

The charts were entirely convincing, but their impact was also underscored by the fact that in his office Hibbeln was wearing the blindingly white uniform of a commander in the U.S. Public Health Service. This was not some crunchy-granola, New Age suggestion for making your life better with natural remedies. This was a public health official explaining that the use of bad fats in the Western diet is a serious public health problem, and that the spread of that diet to other areas of the world is resulting in similar rises in depression and causing a global epidemic.[1]

"The big issue is not only for the individual," Hibbeln said that day in his office. "When one person flies off the handle, somebody is angry. If you have two people fly off the handle, you have a fight. If you have 500, you have a riot. The way the food supply has changed the intake of omega-3 fatty acids affects not only one, not two, but literally millions of people—anyone who buys food."

Hibbeln leaned forward in his chair for emphasis, his arms spread out to indicate the widest possible population. "So what happens to human society when we deprive the brain of these fatty acids generation after generation? What kind of society do we end up with?"

Hibbeln and others who study the role of fatty acids agree with the psychiatric community that much of mental illness—depression in particular—is biochemical. But they contend that the biochemistry of unhappiness is not so much genetic and in need of medication as it is self-inflicted every time we sit down for a meal.

We are what we eat, and apparently what we are is a group of sad people swimming in bad fats.

LESSON ABOUT FAT

Ever since the first humans figured out how to ferment fruit, there has been a connection between what people eat and drink and their mood. People have used this connection to get high, bring on a buzz, hike up their energy, or simply feel better. The chemical compounds in alcohol, caffeine, nicotine, and chocolate are standard mood-altering substances, and there are naturally occurring plants, such as cannabis and poppy, that alter the mind as well, but those mind-bending substances have short-term effects, and their joy and sorrow is the fact that one usually ingests them on purpose for that effect. No one thinks of fat as a mood lifter—or the lack of it as a road to sadness—except that most fats, such as butter and cream, taste great; but fats as chemical compounds are deeply embedded in the way the body operates.

Fat is stored by the body and provides energy when it's needed (unlike carbohydrates, which provide immediate energy). Fats are also the major component of cell walls, and they keep these membranes porous and flexible as well as providing the chemical structure that forms the shape of each cell. In that role, fats affect the overall health of skin, hair, the blood system, and the immune system. Fats also carry vitamins such as A, D, and E. Fats come in all sorts of shapes and sizes such as triglycerides, monoglycerides, fatty acids, and cholesterol. Most of us know about cholesterol because of its presumed connection with heart disease, but the fat story is more complicated, and more significant to overall health, than cholesterol.

Most fats contain some molecules called fatty acids as part of their structure, and these fatty acids come in several forms based on their chemical composition.[2] Some are saturated fatty acids, a description that refers to the number of double chemical bonds and the fact that these fats are saturated with hydrogen atoms. Animal-based foods are full of saturated fatty acids, as those watching their heart or their hips know all too well, but other foods, such as coconut and palm oil, are also full of saturated fatty acids. In contrast, unsaturated fatty acids are missing some hydrogen atoms and they form bonds with other atoms. These unsaturated fats are classified as either monounsaturated or polyunsaturated.

Many people have steered clear of saturated fats for health reasons, but they are unaware that within the polyunsaturated fats is a new threat. Not all polyunsaturated fats are the same, and relying on one kind might be downright dangerous. Omega-3s and omega-6s are the two kinds of polyunsaturated essential fatty acids that are significant in the nutritional model of mental mood. The omega-6s are found in soybean oil, safflower oil, and corn oil, and during hydrogenations they become even more omega-6. These oils, especially in the hydrogenated form, are ubiquitous in processed foods, such as crackers, breakfast cereals, and baked goods, because they are fats that don't go rancid over time. Just pick up any food that comes in a package at the grocery store and look at the ingredients. The list always includes some kind of hydrogenated oil. The omega-3s are also in the foods we eat, but they are much harder to find; foods rich in this type of fatty acid are walnut oil, spinach, and the mother of all omega-3 foods, ocean fish.

Given what we have on the grocery store shelves, and given that most families in Western culture rely on processed food, the current Western diet is chock-full of omega-6s and nearly devoid of omega-3s. As a result, our standard diet is skewed dramatically away from what is nutritionally best for the human mind and body. This is not only the result of a desire for convenience in our meals. In a rush to stave off heart disease, people have rushed to embrace the polyunsaturated fats. We have put down the butter and beef only to take up seed oils and packaged foods packed with hydrogenated soy oil, thinking all polyunsaturated fats are the same. They are not. The body needs both omega-6s and omega-3s, but today we eat way too much of the former and not enough of the latter. And the result, Hibbeln says, shows up dramatically in the brain.

BRAIN FAT

One day more than a decade ago, Joe Hibbeln was working in a research lab when the connection between fats and depression hit him. "I was standing in the lab, holding a human brain in my hand, and it suddenly hit me—the brain is all fat. There's no difference between a stick of butter and your brain. You only need to carve a few lines in the butter and there it is, the brain."

Hibbeln was also aware that fatty acids are integral to cell flexibility in other parts of the body. Therefore, it made sense to suspect that they would have an effect on a fat-rich organ such as the brain.

"All chemical and electrical signals must pass through the outside walls of brain cells," Hibbeln explains. "What's important is that this membrane is composed almost entirely of fats. Membranes at the synapses between brain cells are, in fact, 20 percent essential fatty acids." He briefly describes the brain's intricate workings: "Within this fatty matrix of membranes are ion channels through which electrically charged particles must pass. But the channels are folded into complex and delicate shapes that are crucial to the way they work, including the ability to change shape quickly. As they change shape, the channels either allow the flow of signals or stop them." In other words, the fat composition of this matrix could dramatically affect how the brain thinks, feels, and operates.

"Imagine the difference in trying to operate, say, an eggbeater in lard, margarine, or olive oil," he continues. "The fats all have different physical properties which alter how things, like eggbeaters or proteins, move in them." Hibbeln's description is clear because most of us have had some experience with these ingredients in the kitchen without even acknowledging that they are fats of different consistencies. Obviously, whisking butter is much harder than whisking olive oil.

"The same basic principle applies to the transmission of chemical signals across brain cell synapses," he explains further. "After a neurotransmitter hits a receptor, the receptor must change shape in order to transmit the signal to the inside of the cell." So the composition of the fats holding the ion channels can affect the shape of those channels and influence the way signals are sent throughout the brain. Every time we eat, some of the fats we ingest add their chemical properties to our brains, shifting the ion channels and influencing our brain's signaling.

That's not all. These same fatty acids have been linked to the neurotransmitter serotonin. A neurotransmitter is a chemical that makes the leap between brain cells and enables them to pass signals. As discussed in Chapter 3, serotonin, one of the best-known neurotransmitters, seems to be especially important in mental mood. We know this because antidepressants that increase the level of free floating serotonin in the brain seem to lift the mood of some people who are depressed or feeling low.

Scientists also know that serotonin is important in initial brain development because it signals neurons into their correct locations and assists in the growth of axons and dendrites, the two ends of brain cells that pass electrical signals around. The connection between fatty acids and serotonin comes from several lines of evidence. In 1999 researchers Sylvia de la Pres Owens and Sheila Innes at the University of British Columbia found that 18-day-old piglets whose feed contained omega-3 fatty acids had twice as much serotonin and dopamine in the frontal cortices of their brains as those on standard feed.[3] Research on rats has also shown that doses of omega-3 fatty acids promote neuronic growth, protect cells from destruction due to apoptosis, and make the synapses between the brain cells more efficient.[4] Another study found that monkeys fed omega-3 fatty acids had improved blood flow to the brain, which some believe must affect mood.[5] Again and again, work with lab animals is showing that when they have a diet rich in omega-3 fatty acids, their brains are healthier by all sorts of measures.

Of course, you can't do quite the same sorts of studies with people because it's not ethical or legal to bore into human heads and extract fluid. Fortunately, there's another way to get this information. Hibbeln has found that people with little omega-3 fatty acids in their spinal fluid, which is an indication of how much they have in the brain, also have low levels of a chemical marker for serotonin.[6] He also discovered that people with a history of hostility and violence had low amounts of the marker for serotonin in their spinal fluid, and those low numbers correlated well with low amounts of the omega-3 essential fatty acid called docosahexaenoic acid (DHA).[7]

Taken together, these research results suggest to Hibbeln and others a critical link between mental mood and fatty acid intake.[8] The researchers believe that higher levels of omega-3 fatty acids will allow more blood to the brain, may promote the growth of cells that respond to serotonin, probably help the cells function better and keep them from dying, and make for a more flexible fatty matrix in the brain. In contrast, low levels of omega-3s lead to an imbalance between the essential omega fatty acids in the brain, and this, they believe, has dire consequences.

WEIGHING THE FAT EVIDENCE

When I first began spending time with Joe Hibbeln, I was struck by his enthusiasm. Most scientists are a quiet sort of people, but Joe is more like a house on fire. Every time I sit down with him, he immediately brings out the data charts, one after another, dropping them on my lap or slapping them on a desk and then checking for my reaction. Once when I met him at a Thai restaurant for lunch he brought his portable computer so that we could view data over dessert and coffee. You might expect all these charts and data to cause the eyes to glaze over, especially when one is trying to eat lunch, but they're actually rather mesmerizing because the charts all look the same: one after another shows an upward sweep that indicates the correlation between a lack of omega-3 fatty acids and an increase in negative behaviors and mood. Pick a negative behavior among a group of people, then get the numbers on their fish consumption or levels of omega-3 fatty acids, and there it is—a persistent association between unhappiness and not eating enough food with omega-3 fatty acids.

Specifically, Hibbeln has discovered that rates of depression are associated with how much fish is consumed by people across cultures; when people eat lots of fish they take in a lot of omega-3 fatty acids that balance out the omega-6s. The average New Zealander, for example, eats 40 pounds of fish a year, and 5.8 percent of the population suffers from depression. In Japan, by contrast, where people eat more than 140 pounds of fish on average each year, depression occurs in less than 1 percent of the population. Even when countries with fish-rich diets, like Taiwan and Hong Kong, are removed from the analysis, the correlation holds all over the world.[9] "The rates of bipolar depression and postpartum depression in Iceland are among the lowest in the world," Hibbeln points out. So even though Icelanders may have reason to be blue in such a cold place with long winter nights, they apparently have little seasonal affective disorder.[10] More tellingly, when a Western diet—full of processed and fried foods high in omega-6s and low in omega-3s—is introduced into a culture, as it has been in places such as Greenland and Japan, the rate of depression rises accordingly.

Hibbeln has found the same sort of correlation between the level of seafood consumption and a raft of negative behaviors, including rates of homicide, postpartum depression, hostile behavior, suicide attempts, and bipolar disorder.[11] Recent research by Laure Buydens-Branchy and colleagues has even found a connection between unbalanced essential fatty acids and relapse into cocaine addiction.[12] Even more disturbing, in perpetrators of domestic violence, Hibbeln has found an association between low levels of the DHA type of omega-3 and the presence of a stress hormone called cortisol.[13] Taken together, these studies point to a solid connection between a low level of omega-3 fatty acids and just about every negative, aggressive, and unhappy behavior in the book.

These population studies are further supported by the previously established connection between heart disease and depression. As cardiologists know, people are more likely to have heart disease if they have been diagnosed with depression, and more likely to be depressed if they have had a heart attack. But the connection between depression and heart disease may lie not so much in the heart as in the brain. When heart patients decrease their overall cholesterol by decreasing their fat intake and switching to unsaturated fats, such as margarine and corn oil, Hibbeln says, they actually upset the balance of omega fatty acids and may increase their risk of death by violence, accident, or suicide because their brain fats have been highly skewed. The benefits of omega-3 fatty acids to the treatment of heart disease are so strong that the American Heart Association now recommends that patients on a low-fat diet eat two servings of fish a week, and never mind the fat. That change in diet should elevate mood and improve the entwined relationship between heart and soul.

The connection between omega-3 fatty acids and behavior and mood has been so convincing that clinical trials have been conducted to see if the omega-3s can, in fact, be used to proactively affect the way people feel. In the late 1990s Andrew Stoll of Harvard University conducted a pilot study on patients with bipolar disorder.[14] Half of the 30 subjects in the study were given high doses of the omega-3 fatty acids docosahexaenoic acid (DHA) and eicosapentaenoic acid (EPA) in the form of concentrated fish oil capsules. After four months, those on the omega-3s were doing remarkably better than those on the placebo. "We found that the controls [on the placebo] were much more likely to cycle back into mania or depression while those on fish oil remained stable for a long time. I think these compounds have more of an effect on depression, but it was the mood cycling that was really affected," says Stoll.

Stoll's work has been repeated several times. For example, Boris Nemets and colleagues at Ben Gurion University of the Negev in Israel reported that within two weeks of adding fish oil high in the fatty acid EPA to patients' usual regimen of antidepressant medication, the patients showed marked improvement.[15] By the fourth week, 6 of the 10 taking fish oil, but only 1 in 10 taking the placebo, showed a 50 percent reduction in symptoms as measured in psychological tests. The patients who consumed the fish oil were less depressed, had fewer feelings of guilt and worthlessness, and experienced less insomnia. In a similar study, Malcolm Peet and colleagues at the University of Sheffield in Britain gave 70 depressed patients who had not improved on drugs such as Prozac large doses of omega-3 fatty acids or a placebo for 12 weeks and found that 60 percent of the patients on omega-3s showed marked improvement compared to only 25 percent on the placebo.

Other conditions besides depression and bipolar disorder may also be aided by increases in omega-3 fatty acids. Peet and his group administered omega-3 fatty acids in the form of DHA and EPA to schizophrenic patients who had not improved on medication. Those on the treatment fared much better than those in the control group, who did not get the omega-3s. Furthermore, Peet discovered that EPA in particular worked best for schizophrenics.[16] Jan Mellor and colleagues at the Northern General Hospital in Sheffield also found that schizophrenics given supplements of fatty acids had less severe symptoms, and that the omega-3 fatty acid EPA was especially effective.[17] Schizophrenics apparently benefit from high doses of omega-3 fatty acids because they don't metabolize fatty acids well, and large doses of omega-3s seem to help.

The results of these studies are remarkable because they suggest that schizophrenia can be controlled, even put into remission, with doses of fish oil rather than antipsychotic medication, and currently some clinics are using fish oil as a standard treatment with or without prescription medications.[18]

At the moment, Hibbeln and others are waiting to hear about the results of other ongoing clinical trials using omega-3 fatty acids on various mental conditions. So far, it looks like an increase in omega-3s has promise as a clinical intervention for depression and bipolar disorder, and it may slow the onset of Alzheimer's disease.[19]

The evidence for an association between the importance of essential fatty acids and mental health is becoming so strong that institutions like the American Heart Association and the U.S. Food and Drug Administration are not only taking notice but also making recommendations. They now advise that for a healthy heart and mind adults should eat fatty fish low in mercury twice a week.

The recommendation is the same for pregnant women not only because of the connection between low omega-3s and postpartum depression but also because what women eat during gestation and breast-feeding has a significant impact on infant brain development. Researchers in Norway have discovered that children born to women who took omega-3 supplements during pregnancy and nursing had children with much higher IQs at four years of age.[20] Other clinical trials have shown improvements in infant visual function, growth, and mental development that persist over time when babies receive enough omega-3 fatty acids from their mothers or from infant formula.[21] In response, formula fortified with omega-3 fatty acids is now available in the United States (as it has been in 44 other countries for years).

In fact, there may be an important connection between several generations that were primarily bottle-fed on commercial formula devoid of omega-3 fatty acids followed by a standard child and adult Western diet similarly lacking in these essential fatty acids and negative mood and behavior. In a study of more than 3,000 teens, researchers found an association between low levels of the omega-3 fatty acid DHA and high levels of hostility. And this hostility is not just about teen angst—young adults who eat more fish and other DHA-rich food are simply less angry, cynical, and aggressive.[22]

The studies by credible scientists and the recommendations by credible health agencies all indicate that humans need a good balance of omega-3 and omega-6 fatty acids to develop and maintain a healthy mind. The deeper, broader question is how we got into the predicament of such an unbalanced diet in the first place. If these essential fatty acids are so important to the human species, why are we so lacking in these compounds today? For that answer we have to go way back in history and take a peek at the diet of our ancestors.

THE ANCIENT TABLE

There must have been a time long ago when we were eating better and thinking more cheerfully. Clearly, depression from lack of essential fatty acids is not a good thing, and surely the human lineage has not always been so devoid of the compounds.

Five million years ago, humans shared the forests with apes. At some point our ancestors separated in body type and eventually brain size from our ape cousins, and began to explore resources more far afield. In that evolutionary path, some feel, is the history of the role of omega-3 fatty acids in human behavior and mentality.

According to Michael Crawford of the Institute of Brain Chemistry and Human Nutrition at the University of North London, when humans and apes split, the newly evolving humans moved away from the savannahs and forests of Africa and headed for the coast.[23] As our kind spread across Africa and around the world, we stuck to the rivers, lakes, and beaches, eating all manner of marine life and ingesting large amounts of omega-3 fatty acids.[24] Fish may, in fact, have been the very key to human survival and may even be responsible for the evolution of modern human culture. Loren Cordain of Colorado State University thinks our ancient ancestors may also have ingested these necessary fatty acids on the savannahs as people hunted and butchered meat, digging into bone marrow and eating animal brains, all good sources of omega-3s.[25] No matter the source, it's reasonable to assume that during the time when the large and complex human brain was evolving, the human diet was chock-full of omega-3 fatty acids.[26]

About 200,000 to 100,000 years ago, humans began to look exactly like we do today. Paleontologists classify these fossils as members of our modern genus and species, *Homo sapiens sapiens*. These early people were also dramatically different from their ancestors in terms of culture. Although ancient humans apparently made tools as early as 2.6 million years ago, the more modern people, now equipped with even larger brains, invented art, clothing, jewelry, and new ways to kill. Catching fish was part of this modern way of life. Paleontologists recently discovered that humans lived in Eritrea on the Red Sea and exploited marine life 125,000 years ago.[27] Archaeologists have also found evidence of sophisticated tools at other sites that were used specifically to catch marine life. In Blombos Cave in South Africa, Alison Brooks of George Washington University found carved-bone fishing points that are 70,000 years old. Brooks also found barbed fishing points at a sight in Zaire that may be even older.[28] People used these tools to catch and eat catfish, kob fish, shellfish, and other marine life.

A recent study shows that the bones of humans from 30,000 years ago had the chemical composition of individuals who must have eaten lots of marine life—up to 50 percent of the diet must have come from lakes and oceans, the researchers estimate.[29] These skeletons are radically different in chemical composition from the bones of Neanderthals that lived at the same time and apparently didn't eat much fish. As far as we know, the Neanderthals were not as smart as their fully human cousins and, of course, the Neanderthals died out while we, the fish eaters, kept going.

For 90 percent of human history—tens of thousands of years— we lived this way, moving across the landscape and the beach, consuming more omega-3 fatty acids than today, fueling our brains with a balance of omega-3s and omega-6s.[30] And then about 10,000 years ago humans settled down, often miles away from oceans, rivers, and lakes, and started the agricultural revolution. Gone were the seafood, wild game, and wild plants, replaced by dairy, grain, and domesticated livestock. Up went the omega-6s; down went the omega-3s. The result was a major impact on human health: We developed diseases of civilization, such as cancer, heart disease, and infection.[31]

Much later, after the turn of the twentieth century, the human diet underwent another major change—the industrialization of food. Manufacturers aimed to make food that had a long shelf life, and to do that they utilized barrels of soy, corn, palm, and cottonseed oil, all with huge amounts of omega-6 fatty acids and very little of the omega-3s. Even more recently, manufacturers discovered a way to make processed food "fresh" even longer—by hydrogenating the oil— and that's a problem too. Hydrogenated oil, it turns out, is even more skewed toward omega-6 fatty acids—and today just about everything we eat is full of these bad fats.

The biggest culprit, Hibbeln says, is soy oil. We have been led to believe that soy oil must be good because it's a polyunsaturated fat. Soy oil, however, is very high in omega-6 fatty acids, and unfortunately, our brains are now floating in the stuff. In 1900, according to Hibbeln, the United States produced almost no soybean oil and relied on lard and butter for fats. While recently attending a meeting of the American Oil Chemists' Society, Hibbeln heard that soy oil now accounts for 83 percent of the calories that Americans take in as fat. That means our annual intake of soy oil has increased from 0.02 pounds per person at the turn of the century to 25 pounds a year in 1999—a thousandfold increase. In addition, soybeans and corn are fed to livestock, making their meat also high in omega-6s. And then we fry it all in corn or soy oil, pushing the brain's soup as far as possible from the Paleolithic diet in which it evolved. Today in Western culture, the ratio of omega-6s to omega-3s is 16 to 1.

"These are biologically potent compounds," warns Hibbeln. No part of the body, he maintains, could adjust that fast to that much of a skew, especially the brain, which relies on the right balance of fats to work properly. The result is the epidemic rise in rates of depression and other negative mood disorders.

Although the connection between omega-3 fatty acids and happiness is just being introduced to Western culture, Joe Hibbeln is convinced that many cultures have been on to the idea for a very long time. In a recent article Laura Reis and Hibbeln claim that the fish has been a prominent symbol in many cultures for a reason—because people around the world and throughout human history have understood that regularly eating fish brings peace of mind.[32] Examples range from the use of seafood in traditional Chinese medicine as a way to calm aggressive behavior to the use of fish as a symbol of faith and healing in many religions. Reis and Hibbeln suggest that, consciously or unconsciously, people have made the connection between ingesting a rich source of omega-3 fatty acids and a lift in mood. Fish has therefore become ingrained in our spiritual institutions, sometimes as gods, as well as in folk health practices. Reis and Hibbeln conclude in their paper, "Peace, order, and the reduction of impulsive aggression tend to knit societies together. When paired with religious symbols defining one's cultural identity, fish consumption promotes behaviors that significantly support cultural cohesion."

Fish joins other kinds of foodstuffs that are culturally sanctioned substances to change our mood. Caffeine, for example, is a popular picker-upper in many cultures, including our own. Other cultures prefer to drink tea or chew coca leaves to boost energy and lift mood. And people in affluent societies that are blessed (or cursed) with a wide variety of foods always have something to ingest to make them feel better—chocolate, cookies, something made with heavy cream.

People all over the world routinely use food and drink to lift or otherwise alter their mood. So it should be no surprise that foods can also cause us trouble. Ever since humans gained consciousness, it seems, we have tried to alter that consciousness with what we put in our mouths. And what we put in our mouths is a cultural act that has biological consequences.

CHAPTER 5

STATES OF MIND

Anthropologist Stephen Beckerman is a tall, handsome, lanky man in his early sixties who has spent much of his adult life among Amazonian Indians. Back home he lives outside State College, Pennsylvania, in a giant, someday-we'll-renovate-this, Victorian house. We've gone there to check up on his daughter, who's sick, and after making sure she's okay, we wander up the stairs to Beckerman's home office. The room is full of just the kind of stuff anthropologists should have—spears and woven baskets and stacks of books.

Beckerman hands me a few spears, bows, and arrows, knowing that the anthropologist in me will find these accoutrements of exotic culture irresistible. He explains that these items were made and given to him by the Barí people of Venezuela. Earlier he had mentioned that the Barí are "the sweetest, nicest people on Earth" and claimed he had never, ever, heard any of them raise their voice in anger. Standing in that office full of Amazonian Indian weaponry, it's hard to imagine the peaceful nature of the Barí. But Beckerman knows the difference between peaceful and aggressive people, no matter what their weapons look like. Recently, he started a new research project on another group of Amazonian Indians, the Huaorani of Ecuador, who are said to be angry and fierce, even murderous.[1] Beckerman has found that the Huaorani can be just as self-confident and balanced as the Barí, but they also have the potential to fly into a rage, something a Barí would never do. "The Huaorani never kill in cold blood," Beckerman told me recently, "but they are given to rages if pushed."

Such categorical pronouncements about a generalized cultural personality might seem out of place for someone who is supposed to observe objectively the details of a foreign culture, but anthropologists do this all the time. At a recent academic conference several anthropologists offered up characterizations of the people they study. "Melanesians are so paranoid," said one researcher. "The Hiwi seem to be depressed compared to the Ache," said another. Someone else pointed out that hunter and gatherer groups tend to be "kind, pleasant to be around, and helpful."

Of course, anthropologists aren't the only ones who make sweeping generalizations about the collective personality of particular cultures. Everyone makes such statements, even people who are vehemently opposed to stereotypes. That's because there is some truth to the notion that people from a particular culture frequently act, think, and feel in similar ways and that they can be distinguished as a group from other cultures. In fact, belonging to a culture is all about sharing a particular identity.

How does this happen? Every culture has an ideological framework that produces a collective worldview that in turn colors how its citizens think and behave. It's not so much that people from each place all think exactly alike—even people who live in very homogeneous cultures are individually very different from each other. But a group of people gathered together over time typically develops a set of spoken and unspoken rules that dictate what is normal, what is not, and how to behave around each other.[2] And each group tends to stick to those rules over generations.

Each of us, then, grows up knowing and absorbing the rules of a particular society, and as a result our experience is familiar to others in that culture. We belong to a particular culture that informs our approach to the world, our individual lives, and the way we judge others. It is the lens through which each of us approaches and understands life. Culture is not just how we celebrate holiday rituals or what songs are sung. Culture also frames who fits in and who doesn't, and what the collective does with the individuals who aren't quite "right" according to the cultural rulebook. Every culture defines concepts of mental health and illness and what is accepted as normal or abnormal behavior.

WHAT EXACTLY IS CULTURE?

Anthropologists have argued for years over the definition of culture, and their attempts to define it have not always been all that helpful. "Patterns, explicit and implicit, of and for behavior acquired and transmitted by symbols, constituting the distinctive achievements of human groups, including their embodiments in artifacts," wrote two famous anthropologists 50 years ago after reviewing every definition of culture they could get their hands on.[3] Today, the definitions are just as opaque. "A culture is always a composite, an abstraction created as an analytical simplification."[4] And culture is "everything that people have, think, and do as members of a society."[5] These statements aren't very helpful, or in some cases even understandable.

To complicate matters, ideas about culture change over time. It was very fashionable in the nineteenth century, for example, to rank cultures from primitive to civilized, with Western culture, of course, considered the most sophisticated. At that time scholars also believed that the labels "primitive" and "civilized" were unchangeable—people were not merely born into a culture, they were inherently *of* that culture, and it was impossible to change their rank by moving down or, especially, up the scale. Cultures on the bottom rung were primitive people, and the white males who made up the scale were at the top in the category of civilized people, and that was the natural order of things.

By the beginning of the twentieth century, however, people began to understand that one culture might not be better than another, depending on what you called "culture." Just because Amazonian Indians don't stage operas, does that mean their culture is more primitive than that of industrialized nations? The answer today is a resounding "no," but it took decades for the idea to sink in that every culture is complex and sophisticated in its own right.

We owe the idea of cultural relativity, meaning that all cultures are relatively equal in value and sophistication, in large measure to Franz Boas, an anthropologist who emigrated from Germany to America in 1873. Boas had an eclectic academic background; he was trained in both geology and physics, but ended up following a passion to gather information about cultures threatened with extinction. In North America he found many such groups close at hand.[6] Boas's specialty was Northwest Coast Native Americans, and he spent months among them making notes on their ritual practices. More important, Boas trained a bevy of students, including Margaret Mead, and sent them off around the world to collect information about far-flung societies. The point of their research was to show the West that other ways of living were just as interesting, successful, and decent as those of industrialized societies.[7]

In addition, Boas was strongly motivated to prove that people, as well as cultures, were equal. "As a Jew in 19th century Germany, he also had firsthand experience with the manner in which socially marked groups can come to have their identities or differences inscribed or etched into their makeup and their second-class status assigned to nature," writes anthropologist Jonathan Marks.[8]

When Boas was first working in America, immigration was a controversial topic. Many people wanted to close the borders, believing that immigrants were by nature physically and intellectually inferior to Americans (which is ironic, considering that all of them were descendents of immigrants) and that no amount of good nutrition or good schooling would bring them up to American standards. But Boas set about measuring first- and second-generation immigrants and proved that environment changed biology. In many physical measurements, such as head shape, which is normally very stable within a population, the children of immigrants looked more like typical Americans than their ancestors. Boas proved that "in similar surroundings, people become more physically similar," as Marks says.[9]

Today we know that with good nutrition during development, children of immigrants grow taller than their parents, and that taking on a new culture means more than learning a new language. Bodies and minds change when they leave one culture and enter another.

Most important, Boas and his colleagues were the first to divorce culture from biology and suggest that how we are as cultural beings is not fixed in our genes.[10] Scholars now generally agree that culture is learned and handed down through nonbiological means, that is, you inherit the culture of your ancestors not because it is encoded in your genes but because you are born and grow up in that milieu. Like genes, culture molds our behavior, but even that molding is a dynamic, learned process and not hardwired for any particular culture. That means people can absorb a new culture, follow the ways of two cultures at once, or move across cultures picking up and dropping traditions as they go.

But what is culture? There is the culture with a capital "C" which includes the arts, music, and drama; every group of people has some form of artistic expression that is meaningful given their traditions. Harder to pin down is what we mean by the other kind of culture, the one with the small "c"—the ideology, history, and traditions that influence who we become and how we behave as citizens of one group or another. To decipher this kind of culture, anthropologists start with the most basic difference among human groups: subsistence pattern. For example, Western culture has an industrialized economy while the Australian aborigines were traditionally hunters and gatherers. Other, more subtle differences also define each culture, such as religion, family life, parenting styles, interpersonal interactions, and gift exchange.

EVIDENCE OF CULTURE IN HUMANS AND OTHERS

In essence, culture is that which is learned. But even this definition is problematic, because the very capacity for being a cultural animal is probably genetic, and genes also seem to influence how culture becomes translated in the brain of each person. Another problem with understanding the impact of culture on human lives is that other animals seem to have culture, too. In the 1960s Jane Goodall watched chimpanzees in Africa fashion long sticks to look for termites in ant hills, mash leaves to use as sponges, and wield tree limbs as weapons against baboons.[11] Tool use had traditionally been used as a sign of cultural ability in humans and something that separated humans from apes. Since that time, researchers at every chimpanzee field site across Africa have watched the animals learn to manipulate objects, just as human youngsters take on their culture by playing with toys. In addition, adult chimps display an array of behaviors that are apparently learned and therefore cultural, such as nut cracking and termite fishing.

Recently, researchers from seven long-term chimpanzee field sites got together and made a list of chimpanzee behaviors from the various sites and evaluated them as either universal for all chimps (genetic) or specific to the environment (cultural). They came up with 39 behaviors that are clearly cultural by anyone's definition.[12] Even more interesting, those behaviors vary from group to group. In some areas, for instance, chimpanzees root for ants with a long stick, swipe the ants into a ball, and then flick the ball into their mouths. At other sites the chimps use a short stick and slurp the ants with their tongues. Some chimps clip off the edges of leaves with their teeth as a social display toward others, while chimps at other sites use leaves as napkins. In other words, chimpanzees are not just cultural animals; they

are "multicultural" in the sense that not all members of the same species share the same behaviors—different groups actually make up
their own ways of doing things.[13] And just last year, multicultural
behavior was found in orangutans as well.[14]

We also know that macaque monkeys, with much smaller brains
than apes or humans, can have culture as well. Many years ago a
young female Japanese macaque stunned researchers when she took
a piece of potato and washed it in seawater before eating it. This in-
novative and useful behavior—washing took off all the sand and
added a bit of salt to the potato—then spread among the monkey
troop. Soon all of the animals except the big males were washing their
potatoes.[15] Macaques, then, also demonstrate learned cultural behavior that can disseminate like a new fashion statement.

Even if chimpanzees can fashion tools to get food and wield sticks
to threaten others, and macaques can wash potatoes, we have no idea
how such behavior influences their lives. Chimps do not talk, they do
not actively teach, nor do they use symbols to explain anything to
each other. The observation of cultural acts among nonhuman primates shows that although humans are the only animals to rely on
culture, the evolutionary roots of cultural behavior and thought are
very near the surface.

That said, no one would dispute that humans have moved way
beyond their primate cousins in cultural complexity. We may have
started out chipping rocks, but pretty quickly, in evolutionary terms,
we became masters of culture. If different kinds of tools are any indication of a capacity for culture, we've been at this a very long time.
The fact that early humans made tools suggests that they were able to
think through a problem, such as how to fell game, and come up with
a solution. Then, having picked up this skill, early humans presumably passed it along to their children.

Crude stone tools dating back 2.6 million years have been found
in Ethiopia, apparently made by australopithecines, hominid ancestors with brains the size of a chimp's.[16] As of 1 million years ago, our
ancestors had much larger brains and were making much more re-

fined hand axes and relying on them to kill and butcher game. For all we know, they not only banged and flaked stones to make weapons, they probably also devised and used other artifacts, such as baskets, that simply left no trace in the fossil record. Over the next million years, humans moved out of Africa and across the globe, and they left evidence of their cultural behavior as they went, littering stone tools from Africa to Asia. About 200,000 years ago, people started making art, clothing, and shelters. It's probably no coincidence that human brain size reached its current volume at about the same time as this cultural explosion.[17]

It's hard to say if culture appeared because we needed it to survive, or if culture is just a by-product of an increase in brain size that was selected for other reasons.[18] Some anthropologists believe that our big brains were designed to keep track of social interactions, not to develop sophisticated civilizations.[19] Others have suggested that our brains evolved to help us find particular types of food, care for dependent infants, or accomplish upright walking.[20] There could be many evolutionary explanations for the development and expression of culture in our species, but whatever the reasons for its emergence, culture is now a major player in all we do. We are animals that act out our lives under a certain code, often not even recognizing that code. Unlike other primates, we rely on culture in the deepest psychological sense—in our emotional attachments, social groupings, ways of communicating, how we care for our offspring, and a host of other dimensions of our daily lives. For all these reasons culture as the format of our individual experience is crucial to our behavior and mental outlook.

CULTURE AND THE INDIVIDUAL

Imagine this scene: A woman gives birth at home and when the baby emerges several people—a grandmother, aunts, and female cousins—greet the new baby girl. Very soon after the birth, other visitors start arriving and the mother works hard to keep the baby awake so she will see that many people are happy about her arrival; being friendly and sociable is what life is about, her parents think. For months the baby spends most of her time interacting with everyone who comes by, over and over, and she develops attachments with these people as they look her in the eye, talk to her, and hold her for a while. She sleeps with her mother and is carried all the time, and when she cries, mother is close at hand. But if this little girl cries too much, her mother may consult a special diviner who will make recommendations, such as giving the baby jewelry and coins, to make her happy. At all times the baby is treated with reverence and care because, as everyone knows, this little girl comes from another place, and if she is not happy here in this life, she will go back to her previous spiritual home.

Contrast that scene with this one: A baby arrives at the hospital where she is greeted by a doctor and nurse as well as Mom and Dad. She spends some time on her mother's chest, and then is taken away and put in a nursery for several hours so that her mother can rest. Grandparents arrive at visiting hours, but most of the time the new little baby is either alone in the hospital nursery or with her mother. Her home life for the next few weeks is quiet. When friends and relatives drop by, they are instructed to wash their hands before holding the baby to protect her from germs. Her mother monitors the baby girl's reactions so that these strangers do not overtax her. The baby sleeps alone in a crib in her own room and when she cries, it seems to take a long time for someone to come. Her mother arrives, tired and frazzled, and pats her on the back until she calms down; Mom got this advice for dealing with infant "sleep problems" from the pediatrician, a man who has no children.

While these scenes are about the same event—a birth—they are strikingly different in how those births unfold. The first birth, of a Beng baby in Ivory Coast, West Africa, was witnessed and recorded by anthropologist Alma Gottlieb;[21] the other is a familiar story of Western culture. Both babies come into the world as infant human beings equipped with similar human physiology, and both are fed, clothed, and fussed over. But they have arrived in very different cultures and that will make a major difference in the choices they have in life, their expectations, and what others expect of them. The Beng mother will expect the baby to be part of a social whirl because it has always been like that. The Western mother will expect calm and quiet because that's what infancy was like for her. The Beng child will grow up in a society that grows its own food, and the Western child will, like her parents, grow up and work in an office. As adults each child will perceive her culture as normal and be unconsciously guided by the norms and practices of that culture.

We all learn as children what is acceptable behavior in our culture and what is unacceptable. We might learn, as Beng children do, that interacting with others is the most important thing in life, or we might learn that family comes before all else, or that work is primary. We learn what is polite and what is rude. Each culture also has customs—such as how to talk to strangers or what clothes to wear or how much money to spend—that define behavior from day to day. It is from this extensive network of what is considered normal, acceptable, and customary that each culture stamps behaviors and thoughts as normal or odd, thereby making a standard for judging oneself and others. A community does this through direct instruction, corrective punishment, involvement, and peer pressure. We also watch and learn.

Humans seem to be ready and willing, even compelled, to respond to these measures and follow the rules of a society so that they can belong. We are animals that like to be part of a group. At the same time, buying into one culture, which usually means one set of ways to behave, can be hard for those who by inclination or choice don't fit in. For some people, their own culture is a worrisome place where they just can't adapt, and people who don't naturally fit the status quo can become even more alienated, as we'll see in Chapter 6.

How can we tell what different groups' rules are? As the two birth stories here illustrate, the way to identify the rules of a culture is to compare it to another culture and note how they contrast. In that contrast certain bits of behavior will pop out as interesting, and the idea is to figure out where those bits come from and why they have become integral to that culture. Sometimes these bits have great meaning and profound religious, political, or social significance, and sometimes they are pure whimsy. My brother knew a woman who had an interesting way of making a meatloaf: She would form the loaf of raw meat and then slice off each end before she plopped it in the pan. When my brother asked her why she did it that way, she answered, "This is how I have always done it, the way my mother did it." As far as she was concerned, that's how one makes a meatloaf. When my brother had occasion to ask this woman's mother why she cut off the perfectly good ends of an uncooked meatloaf, the mother replied, "Well, when I mold it the loaf is just too big for the pan I have, so I have to cut some off." Meaningless, but a family tradition nonetheless.

Cultural traditions and values are also hard to understand and trace because they change over time. In just the past few decades smoking has gone from a perfectly ordinary practice to taboo. Homosexuality was recently considered a psychological aberration and now it's an acceptable way of life. In my mother's time, women were supposed to be housewives and now they are expected to contribute to the household economy.

We might think that Western culture changes rapidly because of so many technological innovations—computers, cars, and household appliances have certainly forced this culture to change. But Western culture is not alone in rolling with the cultural punches. I was once discussing Balinese culture with Gigi Weix, an anthropologist at the University of Minnesota who had lived in Bali for some time. I was lamenting all the tourism on the island, assuming the influx of foreigners and foreign money was ruining Balinese traditional culture. "You're wrong about that," Weix said. "The Balinese have always taken what they wanted from other cultures, even when they were occupied by force, and then they have continued on their way." She explained

that the Balinese were thrilled with the current wave of tourism because it has allowed them to pour money into traditional Balinese
arts, supporting a real Renaissance of their traditional high culture,
while the rest of the country's culture was also changing dramatically.
Because of tourism, the Balinese had told her, they could fund dance
schools and support artists and revive *gamelan* music. The Balinese
might now have Coca-Cola and Snickers bars, but they also watch
traditional puppet shows and go to traditional dance lessons.

Cultures are varied and changeable because culture is not a thing,
but the people who act out each culture. And people are always on
the move, which means that culture is never a fixed experience. We
know from studies of immigrants in Western culture that once they
take on Western values, their behaviors, parenting styles, and expectations change.[22] Across the globe there are huge numbers of people
who are right now moving from one culture to another and willing to
learn new ways to think and act.

Because culture is always subject to change and people constantly
flit among cultures, it may seem incongruous that some groups insist
their culture has been around for so many years and that they want to
preserve their heritage. But no culture is frozen in time. American
culture cannot be strictly defined as the way we live now, just as it is
no longer the same culture as the Wild West or of the American Revolution. In the same way, Ache culture cannot be defined as the way
they live now, on missionary reserves, any more than it was fixed in
the recent past when the Aches hunted and gathered in the forests of
Paraguay. We can only describe snapshots of a culture, because it's a
living, almost breathing, aspect of our humanness. It can't be separated from our biology and set apart, as though it were something we
made up or constructed, merely a reflection of our intelligent
thought. Culture is a process, a shared conspiracy to keep groups together for survival and make life more coherent and interesting while
staying alive.

If we didn't have culture, we'd be either bored or dead.

HOW DO WE "GET" CULTURE?

Humans are designed to be cultural animals. Our brains are adapted to learn behavior rather than react in stereotypical ways, to take in stimuli, evaluate them and choose an option, and to strategize. We are born into the world ready and willing to take on culture, to be mentally formed by culture.

How exactly does this happen? How does culture, as a series of visual, auditory, and tactile experiences, shape what a person thinks and feels?

Some anthropologists believe that culture becomes internalized during our upbringing. That is, every cultural sensation is somehow literally etched on the mind, making grooves and lines and thereby affecting the development of personality. Under this scheme, the human mind is a blank slate ready to be imprinted with culture, and a person is solely a cultural construction. But biologists, psychologists, and animal behaviorists find this purely constructed view of a person hard to accept. Studies of temperament, which is one aspect of personality, have shown that we are born reacting a certain way and we keep on reacting in that style forever.[23] Any parent of more than one child can tell you that every child comes out of the womb either kicking and screaming or peaceful as a lamb, and that in later years they react in the same way to life events as teenagers. Research on twins raised apart and on adopted children has also shown that genes influence some behaviors, personality traits, and life choices.[24] In other words, we are each a product of nature and nurture, and we should stop trying to separate these threads or assuming one is of more value or more importance than the other.[25]

In fact, there is evidence that both nature and nurture influence our development. Once an anthropologist visited a psychiatric ward in Japan. A man suffering from schizophrenia lay on his bed, rigid and catatonic, unresponsive to the outside world. But when the anthropologist and his Japanese colleague, a psychiatrist, entered the room, the man sat up, got out of bed, and bowed, and then lay down again in the unresponsive pose.[26] It apparently didn't matter that the patient probably had a genetic predisposition for schizophrenia, and perhaps had a hard life; manners are manners. Despite the profound effect of his illness, which isolated him from society, the patient had retained at least this bit of his learned culture.

The acquisition of culture comes through many channels, and language is one of the most important channels. As children start to talk, listen, and understand the meaning behind words, their world is set into cultural order; with language, the symbolic meaning of a particular culture comes to life. Language, then, is a crucial mechanism by which we become citizens in one culture or another. Indeed, language is a special ability of our species—chimpanzees can be taught to use language symbols, but they do not speak. Language is a magnificent tool for expressing ourselves and interacting with others of our species, and it certainly colors and refracts our worldview. Names for objects, what we call each other, and the way individuals speak to each other about the past and future are all immensely important to how individuals order their world. Words have meaning and they have power, as we know from racist remarks as well from the words of a lover.

Thus, children pick up their culture and the norms of society through words; but actions are also important. For example, Gusii mothers and fathers in western Kenya don't speak much to little children until the children are old enough speak back to them. Babies and young children learn the rules of the Gusii household not by being told, but by watching and by correction.[27] In fact, African children hardly ever speak in the presence of adults because in those cultures children show respect by doing what they are told and not talking about it.[28]

In all these societies we see that language is part of the process of absorbing a culture. It is also extremely important when thinking about mental illness. Can a person be depressed if there is no word for depression? Is someone schizophrenic if they are called a sorcerer instead of crazy? Words reveal how a culture perceives and categorizes, and how people within the culture confront illness, including mental illness. For example, using the words "mental" and "illness" together presupposes that these words belong together and that someone reading or hearing them will understand what is meant. But that might not be true. Perhaps the person reading or hearing the term is a Zuni trance dancer and can't wrap his mind around the term "mental illness" because he just wouldn't ever put those two concepts together.

In addition to language, we absorb culture visually. Western children see cars and roads and airplanes, and think these modes of travel are normal. But a Masai rides in a truck only when he can hitch a ride over mostly dirt roads, and would prefer owning more cattle to owning a car. As human children grow, they observe and construct a worldview based on what they see around them. Every sight, sound, smell, and action is part of cultural experience, and this is closer to what anthropologists really mean by culture.

But the absorption of culture can be a fragile experience. Children growing up in war zones, in times of famine, or in the presence of domestic violence are scarred for life; they certainly have a different worldview from children who grow up feeling safe.[29] For example, work by anthropologist Mark Flinn of the University of Missouri has shown that children never adapt to ongoing household trauma. Their minds and bodies grapple with the stress and express it repeatedly in negative behavior.[30]

Cultures also fall apart when they are conquered and repressed, and the survivors most often become absorbed by the dominant culture. Humans are obviously designed by evolution to continue to absorb culture as adults, whether by choice or by necessity. People who emigrate from one culture to another lose part of their mother culture while taking on a new set of rules and customs. Culture can be both an advantage because it brings us closer together as a group and a burden because it separates groups from each other.

CULTURE AND THE COLLECTIVE MIND

At the age of 33 a woman named Ruth Benedict entered graduate school at Columbia University. It was 1921, and there weren't many women of any age getting advanced degrees, certainly few married women. Benedict, a quiet, reserved person, had been looking for something to give meaning and passion to her life, and she found it in studying anthropology. She also lucked out. Benedict happened to arrive right in the middle of the most glorious time in American anthropology. The discipline was just getting going on this side of the Atlantic, and there were still largely unknown, "exotic" cultures available for study, including many in the United States. Even more fortuitous, Benedict chose to study with the charismatic Franz Boas, who inspired his students—men and women alike—to conduct meaningful fieldwork in far off places.

The best-known anthropologist of that time was Margaret Mead, a fellow student and friend of Ruth Benedict. Mead went on to become a very public anthropologist, the person who brought the words "anthropology" and "culture" into mainstream America with her lectures and popular books.[31] But Benedict worked just as hard and had an even greater influence within the anthropology community. Benedict did fieldwork among the Zuni—photographs show a tall woman in a long dress and stout heels surrounded by Native Americans—and was an expert on trances among Native Americans. She also taught numerous courses and wrote about theoretical issues in a field that was growing by leaps and bounds. Mead may have been the voice and face of early American anthropology, but Benedict was its mind and soul. Together they persuaded many people to think long and hard about how culture influences our minds.

Benedict didn't write a lot, but she had a lot to say and she was able to develop her theories at a relatively leisurely pace. These days professors are driven to "publish or perish," and so we spend hours alone at our computers drafting one journal article after another. The academic culture was different then, and Benedict was able to de-

velop her ideas slowly and with grace. As a result, what she wrote in
the 1920s and 1930s is just as relevant today.

On my desk is a compendium of Benedict's writings, including
poems and letters. The pages have come unglued, but I can pick up
any single sheet and read at random the lilting language that entices
the reader to appreciate cultures other than our own. "Modern civilization, from this point of view, becomes not a necessary pinnacle of
human achievement but one entry in a long series of possible adjust-
ments," she wrote in 1934, admonishing those who think Western
culture is special or superior.[32]

Benedict was especially interested in how culture influences individual psychology; in fact, she believed that a culture could only be
understood on psychological terms. Writing about the various cultures of southwestern Native Americans, including the Zuni, she said,
"It is not only that the understanding of this psychological set is necessary for a descriptive statement of this culture; without it the cultural dynamics of this region are unintelligible."[33]

For Benedict, psychology and culture were entwined, and from
that position she wrote a book called *Patterns of Culture* in which she
compared the ways and beliefs of Zunis, Dobuans from an island off
the coast of New Guinea, and Kwakiutl Indians of the Northwest
Coast. As is now generally accepted, especially among anthropologists, she proposed that every culture has a certain collective personality, and that when you grow up in that culture, or later take on
another culture, you are buying into that collective personality. And
in doing so you join in a sort of contract to act in ways that are acceptable in that culture.

"In all studies of social custom, the crux of the matter is that
the behavior under consideration must pass through the needle's
eye of social acceptance, and only history in its widest sense can
give an account of these social acceptances and rejections," Benedict
wrote in *Patterns of Culture*.[34] The overall personality of a culture
then dictates what is normal and abnormal, and slots individuals
accordingly—in ways that may be very different from what we, in

another culture, think of as normal and abnormal. For example, according to Benedict, a person with seizures was not considered sick among the Shasta Indian community of California, but respected and singled out for leadership positions. In Siberia, shamans were typically neurotic, anxious, and unstable, and these traits were considered a plus.[35] In Dobuan society, where life was rough and no one trusted anyone, a person with an optimistic and happy personality was considered crazy.[36]

Of course, everyone thinks their culture is the best, the most correct, the way people ought to act. "The vast proportion of all individuals who are born into any society always and whatever the idiosyncrasies of its institution, assume, as we have seen, the behavior dictated by that society. This fact is always interpreted by the carriers of that culture as being due to the fact that their particular institutions reflect an ultimate and universal sanity," writes Benedict.[37]

But certain cultural traits that may be considered important at one time can be fostered to the extreme in unhealthy ways. Benedict used our own culture as an example: "Western civilization allows and culturally honors gratifications of the ego which according to any absolute category would be regarded as abnormal. The portrayal of arrogant egoists as family men, as officers of the law, and in business has been a favorite topic of novelists, and they are familiar in every community. Such individuals are probably mentally warped to a greater degree than many inmates of our institutions who are nevertheless socially unavailable." Benedict was able to stand outside her culture and look in, and what she saw was not altogether flattering. In the West we are just as capable of skewing cultural values one way or another and shifting the definitions of normal.

Ruth Benedict, the soft-spoken woman behind the development of American anthropology, pushed hard for people in Western culture to take their blinders off and look around at the array of cultural expression across the globe. She believed that learning about other cultures would lead those in Western culture to realize that there are many other ways to live a life, and thereby understand our own behavior more clearly and perhaps even change some of our accepted patterns or opinions. She underscored the idea that cultures have a certain personality, and that people have to adhere to the behaviors identified by that collective personality to be considered normal, but she also recognized that the definition of normal shifts from one culture to the next.

How exactly does culture refract our assumption of what is normal and abnormal?

CHAPTER 6

RUNNING AMOK IN A BRAIN FOG

ANTHROPOLOGIST JOHN BOCK OF CALIFORNIA STATE UNIVERSITY, Fullerton, once spent three years living in a community in the Okavango Delta of Botswana. By the end of his stay, Bock was fluent in the language and friends with just about everybody, but he soon learned that three years embedded in a culture was not quite enough to feel, think, or act like a native.

One evening Bock was sitting around the fire with a group of friends when an older woman suddenly stood up, ripped off her clothes, and ran screaming into the bush. "Everyone just kept on talking, as if nothing had happened," Bock recalls, his eyes still wide in astonishment. "A few minutes later, one guy excused himself and went after her. We could hear her screaming and running around for about an hour. Then they both came back, perfectly calm, and rejoined the group."

Bock was understandably shaken by this display and when he questioned his friends, they all said the same thing: "Sometimes people have to do that."

For Bock the scene was personally troubling on two counts. First,

as a cultural anthropologist he was there to observe and record behavior, but clearly he had missed something in the previous three
years. He was also quite perturbed to watch someone actually run
amok, a behavior he had never seen, and probably never would see, in
his own culture. In Western culture we occasionally read shocking
news reports of a shooting spree by someone who ran amok, but this
kind of behavior is considered insane, and criminal. We don't seem to
have any counterpart of the more mild form of running about that
Bock witnessed at the campfire. There is no place in Western culture
for ripping off one's clothes and running about screaming in public,
unless you are under three years old, and even then we worry.

What Bock witnessed was a form of psychological expression,
even perhaps an episode of mental illness, different from what we see
in Western culture. His experience is a common one for those who
spend time in a foreign culture. People from different parts of the
globe think, feel, and act differently depending on the culture, and
people go crazy in different ways as well.

CULTURE-BOUND SYNDROMES

Ever since Western explorers began to travel the globe, they recognized that people from other cultures looked and acted differently.
With wonder and fear the explorers wrote in their journals about
behaviors that were unusual, exciting, and exotic.[1] Indeed, much of
what they saw at first contact was so different that no one was really
sure if the seemingly odd behaviors were simply differences between
the savage and the civilized, or real mental illnesses.

By 1904 the famous German psychiatrist Emil Kraepelin introduced a term for such conditions—*vergleichende psychiatrie* or cultural psychiatry—to acknowledge that mental illness can vary among
cultures.[2] In all cases the behavior in question was deemed a mental
illness not just because it was different from Western culture but also
because it was wild or highly unusual and made the observer feel that
the victim must be out of his or her head.

Soon a bevy of anthropologists, students trained during the first few decades of the twentieth century, moved out across the globe with the sole purpose of recording exotic cultures before they were consumed by modernity. These early anthropologists observed any number of extraordinary behaviors that would lead anyone—anyone in Western culture, that is—to wonder about the different guises that mental illness can take, and what exactly constituted mental illness.[3] Clearly, many people in non-Western cultures had mentalities that were often very different from those of Western culture, but they were not insane. Trances, the spirit world, possession by ancestors, voodoo—there was a whole world of ways to think about thinking.

Those descriptions also rocked the foundations of the Western psychiatric community because they brought into our world a very different understanding of what constitutes a healthy or an ill mind. Who's to decide? If a certain act is considered normal in one culture but highly odd in Western culture, does that make it pathological? If a shaman goes into a trance as part of a healing ceremony, is he having a psychotic episode or just doing his job? If a woman in Malaysia screams and jumps away, is she suffering from pathologically high anxiety or is she just highly aware? What if craziness is a reaction to political or cultural change or oppression? Most important, do these behaviors strike us as unusual simply because they are foreign, or are they actually serious disorders and mental diseases?[4]

In other words, just calling a behavior a mental illness does not make it so.

By the 1960s the term "culture-bound syndrome," invented by American-trained Chinese psychiatrist Pow-Ming Yap, had entered the psychiatric literature.[5] The term refers to mental disorders that are seen in one culture and not another. In other words, they are not universal but are bounded by a particular culture. The importance of culture-bound syndromes has also gained prominence for the simple practical reason that immigrants to Western culture (i.e., Europe and North America) are arriving from cultures that are more and more unfamiliar. Immigrants bring their own cosmologies about the mind and body, their own approaches to treatment (what we sometimes call folk remedies), and their own ideas about what it means to be sane or insane.[6] Instead of explorers traveling the globe and "discovering"mental illnesses, the problems have been brought right to the steps of the Western psychiatric clinic.

For Western-trained psychiatrists the situation becomes acute when it appears that some sort of treatment is in order. It may be that those exhibiting what Westerners consider a mental disorder are just part of the normal social fabric of their own culture. Or maybe upon examination the behavior or symptom can indeed be classified by the usual Western terms of anxiety, schizophrenia, depression, and such. In that case the culture-bound syndromes are just another name for the mental illnesses we know, and they can be treated as such. According to anthropologists Ronald Simons and Charles Hughes, psychiatrists might also be presented with conditions that exist in the "twilight zone" of Western psychiatry. These are conditions that don't exactly fit into the lexicon, no matter how you push them. So a psychiatrist may be left wondering if the client should be treated by conventional Western medicine when Western medicine can't even make sense of the patient's experience.

That psychiatric conundrum can also be stood on its head, making the "twilight zone" work both ways. Imagine a group of Shona shamans from Zimbabwe faced with a waiting room of Americans presenting the classic symptoms of depression—lack of sleep, weight loss, lethargy. The shamans might diagnose possession by spirits and recommend special tea and steam baths.[7] Who's to say which diagnosis is correct?

So far, no one has compared diagnoses or remedies in different cultures with any scientific accuracy, so what we label as a mental illness in a culture we know nothing about is presumptuous. A closer analysis is needed to see how that behavior works within the culture in order to determine if it is a mental illness that calls for invasive treatment, a culture-bound pathology that might benefit from a Western intervention, or something that is simply better left alone.

RUNNING AMOK

In the 1700s during his fantastic voyage around the Pacific, Captain Cook and his crew repeatedly witnessed a behavior in Malaysia called *mengamok*, which began with brooding and then erupted into a homicidal frenzy that continued until the perpetrator fell down exhausted. After this display, the victim, called a *pengamok*, slept for a few hours and then woke up remembering nothing.[8] Cook was so taken by this display, and he saw it so often, that he assigned a member of the crew specifically to deal with "amoks." Cook brought his tales back home, and soon the word "amok" entered the English language as a description of someone who runs about creating havoc. By the turn of the century, amok was so well known that it received a psychiatric definition—a real bout of amok must include brooding, random homicide, and subsequent amnesia.[9]

The behavior, or something like it, has been found in various ports across the Pacific, including the Philippines, Singapore, and Indonesia, as well as in Africa and North America. Unleashed craziness and running about is called *cathard* in Polynesia, *mal de pelea* in Puerto Rico, *imu* in Japan, and Whitman syndrome in the United States; although the behavior might not follow the exact trajectory of the Malaysian version, the path is similar enough to group them all as cultural variations of amok.

Amok is considered a mental illness in Western culture, but in other places it has not always had a negative connotation. According to psychiatrist Albert Gaw, the word means "to engage furiously in battle" in Malay, and according to various authors the practice of going amok has deep roots as an appropriate social and political display. Malay warriors actually yelled the word "amok" as they ran at their enemies.[10] Great Indian warriors in the sixteenth century used to burn their property and families right before they swore to die in battle against Portuguese invaders. Hindu warriors also worked themselves into a frenzy before battle, and this particular psychic state was important to the ritual of war.[11] The great naturalist Alfred

Russel Wallace, who spent most of his life in the Pacific, observed that amok behavior was first and foremost a political statement, a way to rage against the practice of slavery and the oppression of colonial masters.[12] In fact, according to Wallace, it was an honorable way to commit suicide as a political statement. According to this view, amok evolved from a political statement to a personal one, from rage against social oppression to individual violence in reaction to frustration, loss, and despair.

In any case, the social context of amok cannot be ignored. The pattern of behavior is not only extreme, it's public; obviously, when a citizen kills people at random, his behavior affects an entire community. In the United States alone there is a long list of men and boys who have taken up arms against people they either didn't know or barely knew. That list includes the My Lai massacre in Vietnam, Charles Whitman climbing a tower at the University of Texas in 1966 and killing 13 people, two boys who shot fellow students at Columbine High School in Colorado in 1999, and any number of (mostly) men who have returned to their former workplaces with shotguns. All of these incidents are public statements about personal anger. The sufferer is most often a young male who has experienced some sort of deep loss or an insult that he finds unacceptable. These males usually have a history of violence and come from unhappy homes, and they are impulsive, immature social misfits.[13]

In many contemporary cases in other countries the young man is separated from his family, working in a foreign environment, or enmeshed in a culture that is undergoing great transition. Canadian psychiatrist Julio Arboleda-Florez points out that running amok is deeply influenced by social factors, such as cultural change, as well as by personal feelings of alienation and a compelling need to assert oneself, albeit in a violent way.[14] As such, running amok is a result of an "overflowing of inner bitterness" against a society that has failed the individual.[15] This explanation is based on the assumption that people, especially men, bottle up their negative feelings until the cork pops and all their anger and frustration are explosively released. Several authors have suggested that this kind of emotional repression accounts for the majority of amok in Malaysia, a very reserved society where it is unacceptable, especially for men, to express resentment, anger, and unhappiness.[16]

Whatever the reasons for each individual's outburst, the appearance and incidence of amok are highly influenced by culture and susceptible to the availability of weaponry and social opportunity. In Laos, for example, when grenades were readily available in the 1960s, there was a rash of men running amok and setting off grenades during a particular religious festival. In 1966, after 20 such attacks, the prime minister of Laos suspended the festival, and the number of amok attacks fell dramatically.[17] Incidents of amok reported almost daily in the newspaper in Malaysia and the Caribbean always include the use of machetes and knives and the hacking off of body parts. It's no surprise that the very definition of amok in the United States includes a gun; this is a society in which guns are culturally acceptable. In fact, in this country how would we know someone was going amok if he didn't have a gun?

Amok as a social condition is recognized when it appears in waves, one unhappy male after another holding hostages or lashing out, and then it disappears from the public eye for a while. Even in Malaysia, the heart of amok country, the incidence dropped dramatically when the central government decided all amok cases would be tried in court. Rates dropped once again with the building of mental hospitals where potential amoks were housed and treated before they could go wild.[18]

Here we have a behavior that was first described as an exotic craziness exhibited only by primitive natives in far-off lands, only to discover that amok is not so exotic and not so crazy. Young men across the globe, in many cultures, have been known to lose control and rail against the forces of oppression and loss, and when they do so, they murder. By definition, real amok—as opposed to the woman who let loose in front of anthropologist John Bock—is a one-time event. After that, the person is locked up, killed, or commits suicide.

Western psychiatrists are presented with a mystery with amok. It's found in so many cultures that it might be biological, a universal phenomenon of human behavior. But it occurs so occasionally that it's hard to pin down a possible biological cause. Some form of epilepsy perhaps? A psychotic break linked to the more familiar diagnosis of schizophrenia? A biochemical imbalance that has yet to be discovered? No one seems very interested, though, in looking for a biological, genetic, or chemical cause. Instead, psychiatrists agree that amok in its many forms is a cultural problem, a mental illness deeply embedded in the relationship between the individual and society.

JUMP UP AND SCREAM

My mother is not a scary person most of the time—a small woman with big hair and a warm smile—but given the right circumstances she can make me scream and jump in terror. The right circumstances are hiccups. I might be talking pleasantly with her, a hiccup here and there, and all of sudden, out of nowhere, she jumps at me and yells, "Boo!" And if that doesn't stop the hiccups—never mind my heart—she'll do it again when I least expect it. No matter how many times I've told her to stop doing this, she does it again—and most of the time, the hiccups go away. I'm left with my eyes wide open and a racing heart, but it works.

My reaction to my mother's sudden attacks is called the "startle reflex." According to Ronald Simons, the startle reflex is universal among mammals; it evolved to make sure we are alert when faced with danger. The price we pay for this handy reflex is that most of the time the threat turns out to be false and we have then spent a lot of energy and worry about nothing. Still, it makes evolutionary sense to have a mechanism in place that reacts swiftly and strongly to sudden, possibly serious threats.

The startle reflex involves any number of physiological systems. The eyes blink, then close and tighten; arms and hands rise up; the stomach clenches; the heart races; and sometimes a high-pitched scream escapes. Then, when the object of the startle is revealed as harmless, the whole body quickly relaxes.[19]

Simons points out that everyone has been startled at some point, and most people remember these experiences. Knowing this, we jump out at friends as a joke and dress up as scary monsters at Halloween. Many people also get a cheap thrill out of being startled, over and over, in horror movies. But an innocent manipulation of the startle reflex can turn ugly if repeated again and again, causing real harm because people never get used to being startled no matter how often it happens and they suffer physiologically and emotionally.

In addition, Simons notes, some people are "hyperstartlers." These poor people, for unknown reasons, react easily to a fright and are often constantly on edge, always waiting for the next surprise.

Simons and others believe that the startle reflex is embedded in culture, part of each society's writing, drama, child play, celebrations, and interpersonal interactions. Nowhere is this connection more interesting than in Malaysia and Indonesia, where some people react so violently, and in such an exaggerated way, that they are considered mentally ill.

The condition is called *latah*, a word that Simons translates as "jumpy" and victims of this condition are called *latahs*. In his fieldwork in Malaysia, Simons found that people with *latah* were common. It was so common that during eight months of fieldwork Simons was able to observe several instances and to conduct interviews with *latahs* and community members who had contact with *latahs*.

Simons discovered that Malaysian *latahs* react like hyperstartlers everywhere—when poked, or in reaction to a loud noise, they jump, drop things, swear, and scream—and they react easily and strongly. But the Malaysian version takes some unfamiliar turns; *latahs* mimic the sounds and body movements of the people around them in either a mirror image or a mocking posture. And when the episode is over, the *latah* often remembers nothing. A Malaysian *latah* is also susceptible to commands given by others during an episode. In fact, ordering a *latah* around—for example, getting them to do something silly like dance or take off their clothes—is considered great entertainment in Malaysia. Oddly enough, it is possible to "catch *latah*"; becoming *latah* sometimes even spreads like an epidemic within and across villages.[20]

On close inspection Simons was able to separate the Malaysian *latah* condition into three types: (1) the basic startle reflex, (2) episodes when the person is imitating others and obeying orders, and (3) episodes when the person seems to be acting a role on purpose. In other words, *latah* is not a straightforward kind of business.

But why in the world do they do it?

Being startled and then imitating others may, some have suggested, be an extension of the type of imitation found in traditional dance and music in South Asia. Or perhaps *latah* is expressed in this region because the South Asian culture provides just the right symbolic template for exaggerated startles.[21] Malaysian *latahs* are most often women, which is not true of hyperstartlers in other cultures, and the gender differences are probably significant. Some of Simons's informants suggested that most of the *latahs* were female because women are weaker and therefore more vulnerable. Others claimed that *latahs* are not born but "bred" through repeated poking and scaring, and that one could, of course, do this more often to women simply because women have little or no power. Or it could be that *latah* is a way for women to reject traditional female roles, or express sexuality that is usually repressed.[22] *Latah* for women would then be a political act, a social rebellion.[23] In that sense, it may be that women become *latah* on purpose to gain power.[24] Or maybe it's all just theater.[25]

Although all these explanations have some value, Simons believes something much more simple is at work. Based on what villagers told him, Simons is convinced that *latah* is how the neurological startle reflex plays out in the Malaysian and Indonesian cultures. In other words, *latah*, as a culturally defined condition, is a whimsical mix of a standard neurological reflex taken in an extreme direction by cultural forces that allow and mold its appearance. Given this definition, it is debatable whether *latah* is a mental illness or just an acceptably odd behavior pattern that gives everybody a good laugh.

Whatever its place in South Asian culture, *latah* is classified in Western psychiatry as a mental illness because it's strange to us, and those afflicted often suffer high anxiety because they can't control their reactions. *Latah* is also classified as a culture-bound syndrome because this specific version occurs more often in Malaysia and Indonesia than in other cultures.[26]

Is it a disease? No one knows. Can it be treated? No one has tried. In the way of human behavior, this condition rises from the foundation of our being a wary mammal, and then moves beyond normalcy into the realm of pathology. As Simons puts it, "The startle reflex is everywhere reflected on and culturally elaborated because it is the response of the body and soul to an overwhelming existential reality. Injury and death wait just around the corner; there are no safe times or places; life is a bubble."[27] The behavior is not really so weird if you think about it. All of us are anxious and jumpy, and for good reason; perhaps some people simply jump more often and more strangely than others.

THE ULTIMATE MALE LOSS

"I think every man has some fear of losing his penis, don't you?" A male colleague posed this question to me as we were standing in the hallway of my department discussing various culture-bound syndromes. I had brought up one that to me sounded really wacky—*koro*—where men think their penis is growing smaller each day, retracting into the body, and ending in the man's death when the penis finally disappears altogether. Not having the genital equipment in question, it was hard for me to answer his question. I have to take on faith that watching one's penis shrink away is a potential fear for all men, and *koro* a manifestation of this universal fear.

Koro is an interesting example of a classic culture-bound syn-

drome because there is no known underlying reflex or biological phenomenon at work. The victim's penis is not shrinking, not retracting
even a millimeter; in other words, *koro* is a purely mind-generated
fear. *Koro* can also spread rapidly from person to person, looking just
like a viral epidemic that has swept through a community. Once it
catches on, everybody has it; during an epidemic, women present with
retraction symptoms too, fearing that their labia or breasts are being
sucked inward.[28] In all cases the victims try various sorts of remedies,
everything from drinking herbal potions and their own urine to putting clamps on their organs to slow the rate of disappearance. Then
just as quickly as it appeared and spread, the condition disappears
and everyone goes back to being normal.

The fear of losing one's genitals has a long history. In the middle
to late 1800s, there were reports of epidemic outbreaks of *koro* in
China and Indonesia. Since the 1960s, the condition has been well
known to Western psychiatrists. Cases have been reported from several non-Asian countries, but it is found most often in South China
and in places with large Chinese communities, such as Taiwan, where
the condition is called *suoyang*.

No one is sure what the word *koro* really means, but it may be
related to the Malay word for tortoise.[29] According to psychiatrist
Wen-Shing Tseng, the head of a tortoise is used in both China and
Malaysia as a symbol of the penis, and so the animal's association
with the symptoms of *koro* is appropriate.[30] The word *koro* can also
be traced to the far-off Indonesian island of Celebes, or Sulawesi,
where there have been ancient reports of a condition called *lasa koro*,
which translates as "shrinking of the genitals."[31]

In the typical case, a man believes that his penis is getting smaller
and smaller, moving into his abdomen, and he knows that once the
penis is gone, he will die. In many cases, especially among non-Asians,
the victim has an underlying psychiatric disorder and a history of
clear mental illness that suggests *koro* is symptomatic of a psychotic
break. Some Western psychiatrists have stated that *koro* is simply fear
of castration made real, but their view is predictably narrow and

Freudian.[32] Others have focused on the fact that the cases are few and far between until there is an outbreak, which suggests some sort of societal anxious reaction.

Wen-Shing Tseng and colleagues looked at two epidemics of *koro* that struck Kwangtung, China, in 1984-1985 and again in 1987 and affected over 2,000 people.[33] This district, it turns out, has a long history of repeated waves of *koro,* and the idea of losing a penis is deeply embedded in the folk beliefs of the culture. The people of Kwangtung believe that ghosts of the dead have no penises and that the ghosts can come back and steal penises from the living. Obviously, it's an easy jump from that line of thought to fear of losing one's genitals, and for that fear to spread across a community like bad gossip. The 1984-1985 epidemic, for example, began with a dire prediction by a fortune-teller that it would be a bad year for everyone. Soon cases of *koro* popped up here and there, and then news of these cases spread, and a year later it was an epidemic. These epidemics are intriguing because the fear is usually fleeting, which doesn't compute as psychosis, or as a deeply rooted fear of castration, or even collective depression. Instead, the incidence and transmission are more in line with other kinds of irrational fears, panic attacks, and intense anxiety over something that can't be stopped.

Collective fear is certainly familiar to most of us. Consider the culture of fear now fostered in Western culture. More people die from auto accidents every year than from terrorist attacks in the United States, but while many people are afraid to fly, very few hesitate to get into a car. And if all your neighbors are fearful, sooner or later you will be scared too.

Wolfgang Jilek and Louise Jilek-Aall of the University of British Columbia looked at all the supposed epidemics of *koro*-like panic, including outbreaks in Thailand and India, and they are convinced that the common denominator is threat. The Jileks contend that people who think their body parts are shrinking are responding to an outside force that is taking away their culture, ethnicity, livelihood, or life. In other words, *koro* is probably an appropriate personal response to forces that threaten one's life or way of life, and thus it makes psychological and evolutionary sense and is not really a disease or mental illness.

Epidemics of *koro* illustrate the role of culture in directing the individual and communal expression of anxiety and panic; because *koro* is all in the mind, the condition must be well known in a culture, and individuals must believe in its power for it to be contagious (although some men reportedly said they had never heard of *koro* before they came down with the condition[34]). In that sense, *koro* is a window into the historically dynamic and plastic nature of mental illness, even in the West.

Psychiatric diseases come in and out of fashion all the time, and in all cultures. For example, high-class Victorians routinely fainted as an escape, but almost nobody does anymore. Today, we accept low levels of anxiety and depression as a part of life. And when a new mental disorder is published, we all wonder if we have it—sex addict, shopaholic, adult ADHD? In other words, we Westerners are just as susceptible to catching a mental illness as the citizens of Kwangtung, China, who can contract *koro* just by hearing that a neighbor is losing his penis.

THE CATALOG

So far, the conditions described here come from the East. I chose them because they are well known in the international psychiatric community, and each has a colorful history. But there are many others. Ever since culture-bound syndromes gained psychiatric legitimacy in the 1960s, anthropologists and psychiatrists have amassed a long list of conditions that seem like a *Ripley's Believe It or Not* of mental illness. Every culture seems to have its own particular manifestations, all with their own exotic names.

Imagine living near the Arctic Circle, where it is always day or always night and the landscape is always white. You might, at times, go crazy and become hysterical, and Arctic people, especially women, often do. One well-known version of Inuit hysteria, called *pibloktog*, begins with days of depression and confusion, then blossoms into behavior such as tearing off clothes and rolling in the snow, eating feces, throwing things, and mimicking others. In the middle of this display, the afflicted women show great physical strength, which makes them hard to restrain. Once the attack is over, they convulse, collapse, weep, and sleep for hours; then they wake up— remembering nothing—and return to normal life.[35] Anthropologist David Landy of the University of Missouri suggests that *pibloktog* might be induced because these women eat entirely too much animal organ meat and fish fat full of vitamin A, which can actually bring on hysterical behavior.[36] Although Landy may be right in some cases, the odd part of Arctic hysteria is that sometimes it happens only once and then disappears.

What about brain fag or brain fog, a condition found in West Africa and most often among young males? The boys complain they can't concentrate, think clearly, or remember anything. There might be blurred vision or a sense of heat or burning about the head. If those boys were growing up in Puerto Rico, they might be said to have a mental illness with the lilting name of *ataque de nervios*. In this case, they would react to a stressful situation by screaming, shaking, and complaining of a feeling of heat rising from the chest. They might faint or have a seizure, and then they would wake clear-headed.[37] Life as a Japanese citizen might bring forth *taijin kyofusho*, which means the victim believes his body parts or bodily functions offend others. Of course in many cultures an evil spirit can enter the body and wreak havoc with the mind. Those possessed lose their identity to the beast, they have no will, and they lose control of their lives.

The list of mental afflictions is extensive, and no one reading about Ghost Sickness among Native Americans who are obsessed with the dead, or Mexican *susto* in which the soul is lost, or *sangue dormido* which produces pain, numbness, stroke, and blindness among the people of the Cape Verde islands can fail to be impressed with the variety of mental maladies that strike humankind. Given the opportunity or the need, humans will spin off in an endless variety of odd behaviors and moods. They will complain of heat or cold, lose track of time and place and body, forget who they are, and even cause harm to themselves and others.

Whisper-thin threads of thought apparently hold our mental state together. And we can mentally break apart in a hundred different ways, all of them entangled in cultural experience.

CULTURE BOUND IN WESTERN CULTURE

The list of Western culture-bound syndromes gathered by psychiatrists, anthropologists, and health care workers includes some very familiar names—anorexia nervosa, bulimia, agoraphobia. These conditions are not listed among the exotic only because they are incorporated into our lives. We see them referred to every morning in the newspaper and talked about on TV at night. These are the culture-bound mental illnesses that we, in Western culture, get to own.

Joan Jacobs Brumberg, professor of history and human development at Cornell University, is just a phone call away. As colleagues and friends we often have lunch together to talk about research, writing, and the ins and outs of being senior women faculty at our home institution. Lucky for me, Brumberg is also an expert on the historical and cultural construction of anorexia nervosa. In her book *Fasting Girls*, Brumberg outlines the history of anorexia from the late nineteenth century to its explosion in the 1980s.[38] As her book persuasively explains, this condition is both biological and cultural, a standard-bearer of a Western culture-bound syndrome.

"Of course anorexia nervosa is a culture-bound syndrome," Brumberg explains as we sit in her house one afternoon, having tea. "We know it is bounded by affluence—it doesn't exist where there isn't a Westernized ideal of beauty with a slim body configuration; it also only appears within a certain kind of family constellation in which the child is not ignored and the parents are there and aware, where the refusal of food is an issue."

Our conversation today is like many Brumberg and I have had. History and anthropology are closely related fields, and we are both interested in the larger social and cultural context in which human behavior is played out.

"It's about food," Brumberg says, stating the obvious. "It's about not eating, and it only works—the behavior only has meaning—in a culture where there is food, because if there isn't food, whatever you don't eat, somebody else will." The condition is also set within a particular mix of social issues. "Anorexia has always been associated with socioeconomic status," Brumberg continues. "Bulimia is much more democratic—it hops around the social class structure in this country. But anorexia is still a rich girl's or rich woman's disorder." Indeed, studies of different immigrant groups show that it takes years or generations of living in affluent Western culture before young women begin to fast; that's why there are so few anorexics from minority groups. And when it pops up in other countries, the girls come from affluent families where food is abundant.

Historically, the rise of anorexia makes sense. "It comes together in the nineteenth century because of the availability of doctors, and of medicine," Brumberg explains. "Daughters of bourgeois families had also lost their economic function. They weren't enjoying the same sort of opportunities as their brothers, and they were worried about whom they might marry and what would happen if they never found a husband. They stayed at home longer without any particular function, waiting to be married." At the same time in history, medicine had entered a new era. "You have doctors who engaged in the process of systematic diagnosis for the first time," Brumberg says. "Why is she emaciated? Is it cancer? Is it consumption? She has lack of appetite—anorexia—but from nervous causes—nervosa." Thus a disease was born.

Anorexia is socially significant because, like *latah* and *koro*, it spreads.[39] "It has this strange contagion factor without a microorganism," says Brumberg. "That has to do with social setting as well." She has documented in photographs and advertisements the reverence for the slim female figure appearing about the same time the condition was named. Media hype for a skinny female figure as the right body fuels the spread of anorexia. If no one saw those ads, if no one owned a scale, would anorexia appear in Western culture at all? The disease's popularity rose in a time and place where there's more food than necessary, as is typical in large-scale agricultural economies, and after the invention of the bathroom scale. In that sense, anorexia is surely culture bound.

Anorexia is not, of course, just a social condition or even a purely psychological phenomenon. Fasting changes the body, unbalances hormones, and who knows what happens biochemically to the brain when there's no nutritional input. Some people probably experience some sort of high after weeks of fasting. Many individuals and groups have certainly used fasting for religious or spiritual purposes; Catholic saints were great fasters, and not eating was the Plains Indian vision quest, for example. For affluent women with anorexia the fasting is not about seeing gods; it's about perfection, control, and, as Brumberg puts it, "using the appetite as voice."[40] The condition in its modern Western form could only come together in a culture where to eat or not is a choice—and therefore a way to express inner demons.

Although it can be difficult to step outside one's own culture and identify mental illnesses that might be culture bound, there are some likely suspects. Agoraphobia, for example, probably occurs only in Western culture, where there are permanent houses and comfortable places to hide. It's hard to imagine being a hunter and gatherer with agoraphobia, or belonging to a nomadic herding culture and having to run for cover whenever the house is moved. Only in sedentary modern cultures are there opportunities to stay at home, as well as reasons to be afraid of going out.

Obesity, some have suggested, may also be a Western culturebound syndrome.[41] If eating too much is the psychological flip side to eating too little, then we in the West certainly have a contagious, debilitating mental disorder on our hands. It may be that ADHD, chronic or situational depression, or panic attacks occur more often in Western culture than in other places because our culture is fertile ground for those particular ills.

Western culture is no different from any other culture in producing conditions that encourage expression of particular psychologies. We cannot assume that *our* mental illnesses are all biological—that is, biochemical—while *their* mental illnesses are all culturally constructed.

THE CULTURE OF THE MIND

Cross-cultural psychiatrists and anthropologists have tried to bring some sort of order to the panoply of weird human behaviors found across the globe. They have come up with the label "culture-bound syndrome" to account for "recurrent, locality-specific patterns of aberrant behavior and troubling experience . . . indigenously considered to be 'illnesses,' or at least afflictions . . . generally limited to specific societies or culture areas."[42] And they have honored those syndromes by placing them in the DSM-IV, which puts the culture-bound syndrome on the Western psychological map.

More important, the label "culture-bound syndrome" has encouraged researchers to look closely at symptoms and make judgments about what a patient is saying and determine if it has psychological weight and might need treatment. The comparison can be enlightening. For example, anthropologist Setha Low has discovered that suffering from nerves in Costa Rica is about the same thing as having nerves as a Guatemalan peasant, a Puerto Rican living in New York City, or someone living in a fishing village in Newfoundland.[43] Everyone with a case of nerves from these disparate locations reports fear, depression, headaches, and pain. While the symptoms aren't exactly the same, they are similar enough to suggest that this mental disorder is not so much culture bound as culture common. The symptoms are close enough to what we call anxiety and depression in the West that we could just as easily use the label "nerves" as a diagnosis for these conditions. In fact, in America we used to say that someone had had a nervous breakdown; although there is no such psychiatric term, the expression was used for decades as a catchall phrase to cover any sort of psychiatric falling-apart, from psychosis to depression. Only recently, as the terms "chronic depression," "psychotic break," and "anxiety attack," among others, have become part of popular American culture and conversation, has the nervous breakdown faded from memory.

Close examination of symptoms and a comparison with a familiar template—most often a Western template—will show that many identified mental illnesses in non-Western cultures are clearly aligned with, or can be fitted into, the familiar Western boxes of depression, anxiety, and psychosis. With that comparison the psychiatrist in Boston confronted with a Costa Rican immigrant complaining of nerves might know where to start.

If we move beyond the Western template for evaluating mental illness in other cultures, the picture becomes more complex, or more interesting, because closely aligned mental experiences can be quite different depending on the social and cultural context of their experience. For example, amok in Malaysia and *pibloktog* (Arctic hysteria) have many of the same symptoms—days of sullenness, exploding craziness, and a psychological crash afterward—but amok adds the horror of homicide, so they are not exactly the same. Depression in some countries includes headaches, while depressives in other countries sleep a lot. Anxiety blossoms in so many ways that it has become the catchall diagnosis here and abroad.

Other negative mental conditions from the catalog of culture-bound syndromes simply don't align with Western expressions at all. *Koro* and *latah* are entirely different, even shocking, to the Western experience. But that street runs two ways; a San hunter and gatherer in Botswana would surely find anorexia nervosa just as bizarre.

We could make a list of all these behaviors, compare, and contrast until we are deaf and blind, and still not know who is mentally ill and who is not, and by which measure we should decide who gets treatment and why. The catalog of culture-bound syndromes is not supposed to be an explanation; it is, in a sense, a celebration of a common humanity. Across the globe and over time, there have always been people who are unhappy, unhinged, anxious, angry, antisocial, and fearful. We know that where we grow up, and who we interact with, molds those emotions, and it is in that translation that we might discover a common, familiar human experience.

CHAPTER 7

CURSED AND HAUNTED

I T'S A GRAY WINTER DAY, AND I'M WANDERING THROUGH THE STREETS OF Oxford, England, looking for the Pitt Rivers Museum, the oldest anthropology museum in the world. After consulting various maps, I finally find it tucked behind the Oxford University Natural History Museum. Turns out, a visitor has to pass among dinosaurs and make a few sharp turns through oddly placed hallways to gain entrance to this most eclectic collection of material goods made by people from around the globe.

General Augustus H. Pitt Rivers was an avid nineteenth-century collector of firearms and other weaponry, but he broadened his tastes to every sort of human endeavor after reading Charles Darwin's *On the Origin of Species*. Pitt Rivers was captivated by the theory of biological evolution, and decided that material culture, too, could be arranged on an evolutionary scale if only one had enough materials at hand. In a twisted vision of how evolution works, Pitt Rivers also thought this scale of material objects would be progressive. That is, he wanted to demonstrate, through objects, that there was a continuum from the simple technology of aboriginal peoples to the more "advanced" artifacts of his contemporary Victorian culture. He was wrong about this, of course, because cultures don't evolve like species and species themselves do not evolve progressively. Nonetheless, we have to be grateful for his enthusiasm because Pitt Rivers amassed one of the greatest collections of human stuff of all time.

The Pitt Rivers Museum contains a raucous jumble of human detritus. Standing in the interior entrance of the museum, I see a large central court, three stories high, and each floor stacked to the rafters with more junk than I had ever seen in my life. It is a wonderland of stuff, a hodgepodge of stuff, stuff so strange and fascinating that I want to shout, "Look at this!" and run from case to case like a three-year-old in a toy store. What also makes the museum so great is that the collections are arranged not by culture but by theme. Boats from all cultures here, measures of time over there, musical instruments on this side, things used for money back there—and instruments of sorcery and magic just over here, if you dare.

People, the Pitt Rivers Museum reveals, are faced with common problems. They want to cross a river, grow plants, get the tangles out of their hair, be entertained, or rid themselves of sorrow, for example. The solutions to these problems are local, which means they are highly influenced by a specific culture built from tradition, available materials, and previous knowledge.

In awe of this carnival of human junk, I wander from case to case, marveling at what people can do with a little imagination and a bit of string or cloth. I spend a long time staring at the case labeled "measures of time" and am just as fascinated by the hair combs. Then I round the corner and come face to face with a glass case full of shrunken heads. Thinking a closer look will be intellectually stimulating, I step up and peer at tiny dried-up black heads with bits of hair sticking out here and there. So far, the museum has been really fun, like meandering through a cross-cultural garage sale. But standing in front of those shrunken heads is, well, scary. Their eyes and mouths are sewn shut, their skulls have been removed, and they have been tanned like leather for a sofa. Yet the individual facial features are intact; these were real people, really shrunken, and really creepy.

As I walk quickly away from the case, it occurs to me that those heads represented the darkest side of human behavior, and are scary to me because such darkness is unfamiliar. Combing one's hair with a piece of shell is a lovely act, and one that most of us can relate to because we need to untangle our hair. But to take the head of an enemy and preserve it as a talisman, a magic object, represents an outlook very unfamiliar to Western culture, and therefore is scary.

In Western culture we don't believe much in curses, trances, the evil eye, voodoo, sorcery, and magic. Nor do we believe that someone can make us crazy or bend our mind against our will.

But maybe we should.

BLAMING THE GHOSTS AND SPIRITS

I do not believe in ghosts and spirits, or at least I don't think so. I was brought up Catholic and used to believe sincerely in dead saints as magical creatures who could bring me a new bicycle if I prayed hard enough and long enough. I lost those beliefs as I grew up, and these days I would maintain that I don't have any spiritual beliefs at all. But I also have to admit that something happens to me when I spend time in a foreign country—I suddenly become a believer.

For example, I once spent several months doing research at a series of Balinese temples. I was there to watch the temple monkeys and their interactions with tourists, but it's impossible to sit at a temple all day and not absorb some spirituality, especially in Bali, where religion and belief are doled out with grace. Every morning I would watch women from the village come quietly to the temple. Each had a beaded basket full of flowers and fruit on her head, and a thin wisp of incense smoke rose from these offerings. The women would glide up to the stone steps of the temple and crouch down, the baskets still balanced on their heads. It seemed as if their ballet-like movements came not just from training but also from some inner spiritual place that compelled them, like saints, to appease the gods. They left behind tiny palm mats with pieces of fruit and burning incense on temple walls and gate entries to ward off the evil spirits, and at one temple the offerings were placed at the root of a great tree. In all cases, the monkeys soon rushed out of the forest and gobbled up the fruit and flowers, which seemed quite impolite to me. When I asked one of my Balinese friends about this, she laughed and explained, "It's fine for the monkeys to eat the offerings. After all, they might be spirits, too, and so the offering is for them if they want it." After a few months of this morning ritual, I, too, seemed to accept that it was normal behavior to placate the gods each morning and I felt deprived when I returned to Western culture where no one did anything to keep the evil spirits away.

Then one day I, an agnostic, bought a Balinese angel because I was told she is the Balinese baby god who protects children; that angel now flies over my daughter's bed. Other Balinese angels hang from the corners of my living room, and I joke that they are keeping the ceiling up, but in truth I am never quite comfortable when we take them down to paint or renovate or move. I also have an Australian aboriginal painting of kangaroos leaping in the cosmos, and I feel the spirituality of that painting more deeply than I feel the presence of God in a Western church. Apparently, I have trouble accepting Western religious practices, but I easily embrace spiritual rituals and objects from other cultures.

I have also had an experience with a ghost. When my father died several years ago, I swear that he stood at my bedroom door and checked on me right after he passed away.

The point is, even a confessed nonbeliever can believe in spirits and ghosts when the conditions are right. My beliefs are positive, harmless, and personal, but around the world, ghosts and spirits are most often psychologically malevolent, and in their bad behavior they serve a social purpose.

Myanmar, or Burma, is one of the more exotic societies because it is virtually unknown to the West. Set next to India, to the west of

Thailand, it is an ancient civilization of great beauty and history. These days the gates of Myanmar are closed to all but the most rugged tourists, and those who do go report that the country is stalled in the 1930s, a peasant culture without much technology or global anything. Information is not going into the country and very little comes out. But in the 1960s anthropologist Melford Spiro conducted the classic study of supernaturally caused illnesses, especially mental illness, in several villages outside the capital, then called Mandalay.[1]

Spiro discovered that in Burma there are both witches and ghosts that cause trouble. Witches are distinguished by gender; female witches do their own dirty work, while male witches can be hired by anyone to send out ghosts and cause harm. Male witches might use evil spirits called *nats*, which are conjured up with bits of animal parts combined with this and that in a concoction that is set afire at a shrine where the *nat* lives. The witch then controls the *nat* and he can send it to hurt others.

Female witches are more common, and they are responsible for ailments like intestinal trouble and eye infections as well as mental illness. They bring on insanity by putting a foreign object in the victim's food—the diagnosis is insanity by poisoning. Mental illness also occurs when the particular supernatural ghosts called "Thirty-Seven *Nats*" circle in and take control. These ghosts died violent deaths and it's a good idea to appease them with gifts of food, or by wearing an amulet, or with dancing and singing. But once in charge, the Thirty-Seven *Nats* can cause a person to become unconscious or violent, yell obscenities, or go into a trance. Women are much more susceptible to these supernatural beings than men.

Spiro explains that encounters with ghosts and spirits are considered normal for the Burmese, but when an encounter causes mental and physical symptoms, then the spirit needs attention. Sometimes the possession occurs when the victim is unconscious, or in a trance, and sometimes it happens in broad daylight. The afflicted can get relief by consulting a woman shaman who has married a *nat* herself, or a male Buddhist exorcist. The chosen health professional proceeds accordingly. If the illness is witch caused, there's lots of finger pointing at witches in the village or in neighboring villages. If the cause is a *nat*, witches might become possessed themselves and take on the spirit and exorcise it with a séance, or shamans will go into a trance and fight the *nat* by proxy. In all cases the healer is trying to get rid of an outside force causing the symptoms; the individual is seen as a passive agent. Unlike Western psychiatry, in which the patient's mind and behavior are at issue, in the Burmese cosmology it's nobody's fault but the supernatural forces.[2]

A BELIEF IN POSSESSION AS A CAUSE of mental illness is common across the globe.[3] Possession has been named as the cause of distress in cases of *amok*, *latah*, *pibloktog*, and many other conditions or behaviors that Western culture would consider mental illnesses. There are as many kinds of possessing spirits as there are cultures. For example, over a two-year period, psychiatrist Sangun Suwanlert interviewed 90 people who had been possessed in rural Thailand.[4] The ghosts of dead ancestors possessed some and made them crazy, others were possessed by spirits that had risen from sacred objects, and still others had been invaded by spirits of the living called *phi pob*. According to Spiro's subjects, the etiology of the demon makes a huge difference; a spirit from heaven is good and likes to have fun, an ancestor spirit is defiant and looks everyone in the eye, a spirit from an object displays quiet superiority.[5] These spirits can come and go at will, bringing on symptoms and then allowing the victim to lapse into remission. Sometimes, particularly in the case of spirits of the living—the *phi pob*—the possession is temporary and once gone never returns.

Across cultures, possession is described as an alien force that enters the mind. Those possessed report a loss of memory and consciousness, a loss of identity, and a loss of control. They express physical symptoms, such as shaking, falling, sensitivity to pain, and voice tone changes, and they often go into some sort of trance. Possession happens more often to women and, cross-cultural psychiatrists have discovered, most often to the socially vulnerable—those of low socioeconomic status and little education.[6]

The best way to cure the symptoms of possession is, obviously, to get rid of whatever has taken control. In Burma the sufferer undergoes treatment under the guidance of a healer; he or she goes into a trance during an exorcism and speaks the words of the spirit. Spiro suggests that the trance is an opportunity for the victim to release negative unconscious thoughts and behaviors, and to resolve any psychological conflicts in a safe venue. The victim is buoyed up by the healer's power and ability to bring in the good Buddhist forces to battle the bad anti-Buddhist evil forces. Family and friends are also sources of comfort. As Spiro writes, "Since the patient's abnormal behavior, no matter how antisocial, is attributed to supernatural causes, he is absolved of all blame and, instead, offered every form of sympathy and emotional support."[7] On the other hand, the possessed might get rid of a spirit by passing it along to someone else, as can happen with *phi pob* in rural Thailand.[8]

The idea of possession as a cause of insanity is familiar to Western culture as well; there have always been spirits and demons out there ready to jump into our brains. From the Middle Ages in Europe to witch hunts in colonial America we, too, have had our cast of malevolent characters.[9] And possession has long been connected with various Western religions, for example, when God enters the body of a saint. What is more interesting is how these possessions played out socially. Sometimes the victim was shunned and feared, and at other times they were innocents playing host to something that might be good or evil.

Possession as an explanation for odd behavior doesn't really fit with any modern Western explanation of mental disorders. From a psychiatric point of view, people claiming possession are considered psychotic—that is, really out of their minds and in need of serious intervention. When faced with possession in other cultures, the Western psychiatric diagnosis is a "dissociative trance disorder" under the category "dissociative disorder not otherwise specified,"[10] and the label is "psychosis." The person should be treated with medication, social services, and perhaps talk therapy, if possible. Where other cultures see possession, we see mental illness that must have a different etiology. Yet those claiming possession don't always follow the normal path of psychosis familiar to Westerners. In fact, the possessing spirits that make people crazy often come and go, they can be ritually exorcised, and sometimes they leave forever. So possession itself as described in non-Western cultures is not necessarily a mental illness in the Western way of thinking, but it is an explanation for what we would call mental illness.

The point in non-Western cultures is to lay blame. Nicely enough, the blame is not on the victim. Instead, a spirit has entered the body and this is no one's fault. In that sense the diagnosis is social, not personal. The victim has fallen prey to some evil and the possession is acknowledged publicly because the person is not at fault; victims need not feel shame because they were possessed without their consent. And the treatment is exorcism, which apparently has excellent therapeutic value. Possession as a mental illness, then, is a cultural and social experience that is also treated socially.

WHO SENDS THE DEMONS?

In most cultures where possession is possible, witches and sorcerers guide the bad spirits toward intended victims. Anthropologists differentiate witchcraft from sorcery by the motivation behind the spells. Witches don't always know they are witches, and even when they know, they are compelled by their craft to cause harm and are not always responsible for their actions. Sorcerers, on the other hand, are usually trained in their arts, and their spells are typically targeted at particular people for particular reasons. Sorcerers also differ from witches in that they use various accoutrements, while witches usually cast their spells just by thinking them.[11]

British anthropologist E. E. Evans-Pritchard came up with this division between witchcraft and sorcery, which is now widely accepted, after he spent time in the 1920s living with the Azande people of the Sudan. The Azande, Evans-Pritchard discovered, ran their lives through a complex system of witchcraft and magic that infiltrated every move, every day, by every citizen of the community. "I had no difficulty in discovering what Azande think about witchcraft, nor observing what they do to combat it," he wrote in his famous ethnography *Witchcraft, Oracles and Magic Among the Azande.* "These ideas and actions are on the surface of their life and are accessible to anyone who lives for a few weeks in their homesteads. Every Zande is an authority on witchcraft."[12]

The ability to be an Azande witch lies in the intestines, they believe. "It is attached to the edge of the liver," one informant told Evans-Pritchard. "When people cut open the belly they have only to pierce it and witchcraft-substance bursts through with a pop."[13] Having this special intestinal substance allows the witch to send out spells to unsuspecting victims. Azande witches "perform no rite, utter no spell, and possess no medicine," according to Evans-Pritchard.[14] Instead, Azande witchcraft is performed in the realm of the mind, where witches send out spells and victims mentally receive them. The Azande also claim that the intestinal witchcraft substance allows the soul of a witch to go out at night and harm others. Witchness is inherited along gender lines, passed down from mother to daughter or father to son. But in the way of all cultures, the rules of inheritance are often bent this way and that to clear an accused witch, or explain how someone else must be a witch although there are no witches in his or her family.

The only real way to identify a witch is to consult an oracle. The oracle, in the Azande case, is a ritual that involves poisoning two chickens and watching their reaction. When something bad happens, the victim obtains special plants, makes a paste, and forces it down the chickens' throats. How the chickens die is a message about who sent the curse and how to find relief. These oracles can also predict the future and be useful in skirting bad magic. "There is nothing that pleases him more than to spend a morning in the bush with a basketful of chickens and a good supply of poison, making detailed inquiry about his health, the welfare of his family, and the auspiciousness of his undertaking," quips Evans-Pritchard.[15]

Suspected witchcraft is embedded in all the negative aspects of Azande life, such as death, illness, injury, or as warnings about bad things to come. Even so, the Azande don't think of witchcraft as something bad that has been aimed at a particular person for a specific reason, but as a generalized explanation for whatever happens. In that sense, as Evans-Pritchard puts it, witchcraft is a "natural philosophy" that weaves together the unexplainable into a comprehensive whole.[16] Witchcraft is a way to order society, conduct human affairs, and solve problems; it is a moral value for the Azande.

"If blight seizes the ground-nut crop it is witchcraft; if the bush is vainly scoured for game it is witchcraft; if termites do not rise when their swarming is due and a cold, useless night is spent in waiting for their flight, it is witchcraft; if a wife is sulky and unresponsive to her husband it is witchcraft; if a prince is cold and distant with his subjects it is witchcraft; if a magical rite fails to achieve its purpose, it is witchcraft; if, in fact, any failure or misfortune falls upon anyone at any time and in any relation to any of the manifold activities of his life it may be due to witchcraft," wrote Evans-Pritchard.[17] While the Azande understand that sickness causes death and falling might cause a broken arm, at a deeper level they reason that witchcraft brought on the illness or pushed the person over. At the same time, no Azande lives in constant terror of witches because as they know life is full of misfortune. In that sense, witchcraft brings an order to everyday life among the Azande because it provides explanations to the age-old question, "Why me?"

Witchcraft also provides a way to fight against bad events in life; if witches are to blame, maybe they can be stopped. Most often, enemies cast a spell for the usual reasons—jealousy, anger, greed—and those close at hand are the most likely witches at fault. If no one comes readily to mind, the victim widens the circle, goes back in time, and casts about until someone pops up as a likely suspect. There are a million ways to connect people along paths of deeds and misdeeds, a million ways to assume that someone hates you enough to cause harm. In that sense, witchcraft is also a way to understand Azande interpersonal relationships and how they are perceived, manipulated, and built over generations. And that history is constantly in motion, as it always is among people. The author of today's misfortune will be a non-witch tomorrow, and it is just as likely that a friend will be a witch at the next opportunity. In the litany of who sent which curse and when is the history of Azande social interaction.

Once a witch is named, the confrontation is public, and so the social process of witchcraft reinforces communal notions of good and bad, and exercises the moral fiber of the culture. Members of the community are not just worried about receiving spells, they are worried about the accusation of witchcraft. Thus, Azande witchcraft is a cosmology that keeps people in line; it is, in all its dynamic glory, a moral code.

Azande witchcraft is not a matter of evil versus good, as it is conceived in Western culture, but a way to make sense of the unexplainable, to recognize malevolence in a very public way and get unpleasantness into the public domain where it can be addressed and thereby erased. But in most other cultures, witchcraft is considered evil and witches are feared because they always cause harm. In most places witches are motivated by uncontrollable impulses, full of hate, and able to act in superhuman ways. For example, in some cultures witches walk upside down, or eat human flesh, or come out only at night. Their curses are often random, and usually hard to track down because witches perform their spells alone, from mind to mind, without the aid of props. Victims of spells, curses, and magic can be deeply affected; physically and mentally they are greatly stressed and, of course, the more one believes, the stronger the effect. Witches and sorcerers are not considered insane in cultures where they practice, but they are the source of what might become insanity in their victims. Witches and sorcerers are powerful people full of magic, and therefore not to be taken lightly.

MENTAL HEALERS

If witches and sorcerers are to blame for possession, who rids the mind of their spells? In most cultures shamans—men and sometimes women connected to the spirit world—play the role of healer and are therefore powerful members of their community. The shaman is a religious leader, who either trained for the job or followed in the footsteps of a parent. Shamans might also be people just on the edge of reality who slip easily into the role because their particular experience of reality is different—they are more "out there" than others in the community.

Shamans usually practice their art by going into a trance, often with the aid of mood-altering drugs, and then calling a spirit for help. Together, the shaman and spirit bring relief, and in that act the shaman becomes part spirit as well.[18] Sometimes the shaman just divines (diagnoses) the problem and then it is for others to figure out what to do. In all cases shamans have secret knowledge from other worlds, and they are the medium through which these spirits speak and act. The shamanistic role also includes passing knowledge among communities, and so the shaman is the keeper of cultural mythology and symbolism.

Shamans are sometimes drawn to the profession because they have certain tendencies. Those who are close to the spirit world, that is, not anchored well in reality, are good candidates because they seem to be halfway into another universe and primed to communicate across boundaries. They are most often those who have a history of psychological instability.[19] For example, the Tungus people of Siberia are known for a high rate of Arctic hysteria and shamans are most often those with a history of psychological instability. According to Tungus theory, spirits are looking for a body, and although the spirit might initially come to someone against his or her will, eventually the shaman will tame the spirit and control it. Reports from early fieldwork note that a Tungus shaman wears a colorful costume, flashes mirrors, and beats a tambourine to bring in the spirit. His or her job is in the realm of the soul—by catching a departing soul, for example, death can be stopped. In fact, the word "shaman" is a Tungus word brought to us from Russia in the 1600s, and since then used repeatedly in descriptions of Arctic and North American populations.[20]

In many ways shamans are like the powerful magical operators of other cosmologies.[21] They are the saints who talk to God or the believers who are cured by God's touch. Shaman is also a rather fluid description, used loosely these days to refer to anyone who has a special role in a culture as a healer, diviner, or prophet who interacts with spirits.

Shamans are held in high regard in indigenous communities, but their thoughts and behaviors would probably place them in a psychiatric ward in Western culture—shamanistic claims that they see and feel in another reality sound very like schizophrenia. Anthropologists have noted that shamans in indigenous cultures are indeed akin to schizophrenics in Western culture, but the difference in how that mentality plays out lies in the way each society responds.[22] Shamans and schizophrenics both progress through the same psychological process that includes disturbed thinking, intense emotionality, and wild behavior. But where we in the West see crazy people, other cultures see opportunity. Those entering a new reality and talking to spirits are considered to be on a path of enlightenment, journeying into the spirit world where they can be of help. In the long run the shamans probably do better in terms of life because they have a role in society. In contrast, a schizophrenic in the West is most often caught in a downward spiral of psychiatric care and completely divorced from society because his or her connection to another reality is considered a diagnostic symptom of mental illness, not an advantage.

ANOTHER SELF

On Saturday mornings my daughter and I get up late and fire up the stereo. Still in our pajamas, teeth not brushed, hair sticking up from a night on the pillow, we begin to dance. As the music pounds through the house, we whirl around, happy not to be going to work or school, happy to be together dancing and singing. We lose ourselves for a while, heads back, arms out, yelling and spinning. And it feels great. But after a few songs, I get worn out and turn down the music and nudge my little girl toward breakfast, back to "normal" life.

Our dancing is all in fun, and it doesn't go any further because neither of us are shamans. We don't have the power or the status, or the physical and mental ability, to transport ourselves to a new reality. Some might suggest we are more anchored in life, mentally stable, but others might suggest that we are lacking because, as Westerners, we can't move from this reality to the next.

When anthropologist Richard Lee of the University of Toronto began observing the Dobe !Kung, sometimes known as Bushmen, in Botswana in the 1960s, they were still making a living as hunters and gatherers. Since that time, the !Kung have become settled, Westernized, and have given up most of their hunting and gathering ways. Lee and others were lucky enough to observe and record the details of their lifestyle and belief system before Westernization had too much influence. Hunters and gatherers are important to anthropologists not only because their lives are so different from our industrialized ways but also because the human career began in hunting and gathering. In fact, for 95 percent of human history we made a living by gathering plants and hunting small animals. So anthropologists have been interested in how people today gather and hunt for their food, how their interpersonal relationships are conducted, and how their belief systems mold the culture as a window to our not-so-ancient past.

According to Lee, !Kung cosmology includes various gods as well as animal spirits and ghosts.[23] These ghosts, called //gangwasi (the // denote clicks in the mouth), are spirits of the recently dead that hang around just outside the village and cause all sorts of illness and misfortune. In response, the !Kung have a host of spells, curses, and potions to banish the ghosts and cure the sick or harmed.

They also dance.

The dance begins in the evening as women form a tight circle around a campfire and begin to sing. Then the men start to dance. The !Kung claim there is a substance called *n/um* in the belly of some men and women, and during a dance, as the body heats up, the *n/um* boils and rises until it reaches the brain and explodes. At that point the trance dancer is filled with energy and power. "You feel your blood become very hot, just like blood boiling on a fire, and then you start healing," says one of Lee's informants.[24] A person in a trance is easy to spot—heavy breathing, blank stares, sweating, and even more intense dancing—and is now in what the !Kung call !*kai*, or half-death. Eventually, the individual collapses, and then rises, still in a trance, to move about, touching everyone and sharing the power and energy by rubbing sweat on their skin. In this way the healer is able to instill some *n/um* into the afflicted, talk to the ghosts, and heal what has gone wrong. All through the night, dancers reach !*kai* and one by one they move about the audience, sharing the responsibility of healing.[25] Then they fall asleep, exhausted from the dancing and the healing.

What makes !Kung trance dancing unusual is that the community shares the experience. Every man wants to grow up to be a healer, and women can also have the job; about half the men and a third of the women in the Dobe community were healers when Lee was observing the dances. Depending on the size of the community, dances occur all the time; large communities might have dances several nights a week.

As Lee points out, the healing trances are just as likely to work as any kind of medicine. Ninety percent of illnesses disappear by themselves, so the healing trance is almost as effective as going to the doctor's office. It's certainly more fun. Women sing, men dance (or the reverse if it's a women's dance), and everyone sits around and chats. The party, which goes long into the night, is a community gathering with a purpose—to banish the bad ghosts.

At the heart of the healing is the fact that many people in the community dance until they are out of their minds. Remarkably, from a Western point of view, dancers work themselves into a state of altered consciousness without the aid of any kind of chemicals, and when the healing work is done they come back into this world without a hitch. Healers are seen as important members of the community, not crazy people, and the act of going to a different space is done for the common good.

Lee points out that trance dancing is one way that the !Kung reinforce their sense of community, of sharing. As hunters and gatherers, all food is shared—sharing resources has always been the key to hunter and gatherer survival. The compelling need to share is repeated in many !Kung rituals and beliefs. Illness and harm come not from one another but from the malevolent ghosts just outside the compound. Together the many healers, the singers, and all those in attendance banish the ghosts during the trance dance. As Lee puts it, "The Bushmen, simply by attributing misfortune to an external source, have evolved a protective system that dissipates, rather than intensifies, interpersonal history."[26]

We don't have much trance dancing in Western culture, but there are social outlets that allow Westerners to participate at very deep spiritual levels; we call most of these events religious, or spiritual, or just wild parties. In fact, one study found that in 90 percent of 488 cultures sampled there was some form of altered state that was completely acceptable to the culture, and alcohol consumption wasn't even included in the analysis.[27] If alcohol and recreational drugs are included, it looks like every culture on Earth allows citizens socially sanctioned time to be out of their minds.

These altered states are not mental illness but a collective vacation from reality. The ability to be someplace else mentally, and the idea that culture helps us get there may, in fact, be part of human nature. It may be that when humans reached a point of consciousness in which we were able to remember the past, think about the future, and worry, we invented ways to sidestep the fear that bubbled up in our conscious brains. Lizards and mice don't go into trances, but then maybe they don't have to. People dance all night and enter another level of consciousness because the human mind is able to easily reach that plane. And maybe altered consciousness is a human necessity.

Australian aborigines speak of the dreaming, a cosmology so alien to the Western thought process that no one seems to be able to explain it adequately to our one-dimensional concept of life. For many aboriginal communities the universe is not just the here and now but a mixture of past and present, before and after, here and there; every object and scene has yet another layer. To them every inch of the landscape has a spiritual side and one simply has to look correctly, or perceive in a certain way, to see those spirits.

Most non-Western cultures, in fact, have a cosmology that includes a spiritual level, and believe in a dark side, such as voodoo or magic, that accounts for the bad things in life. Unfortunately, we in the West have nothing so useful to explain what happens or how we think, except for those who blame the Devil. Instead, we cling to a psychological cosmology that can be perceived only by the senses of smell, sight, touch, and hearing, and in that system there is very little room for other senses of reality. We blame biology, not the spirits, and we look to pills and therapists to help us, not witch doctors. We may feel smug about being modern, biologically sophisticated, and scientific, but blaming biology for mental illnesses is a belief system, too. We don't point the finger at witches as the cause of our misery, nor turn to shamans to bring on the spirits for a cure, but instead look to medical science to diagnose, cure, and calm the mind. Yet our choice of cure doesn't seem to be all that successful if the growing epidemic of mental illness is any indication.

Like any good tribe, we have collectively agreed that genes, physiology, biochemistry, and biology are where the answers must lie, even though no one has really demonstrated that our belief system, and our treatments, are any better than poisoning chickens, casting a spell, or trance dancing.

CHAPTER 8

A HAPPY ENDING

MY QUEST FOR A FULLER UNDERSTANDING OF MENTAL DISORDERS eventually leads me to Arthur Kleinman. Kleinman is a psychiatrist and medical anthropologist interested in how medical practice plays out across cultures, including Western culture. He is most interested in mental illness and other kinds of human suffering that bring people to their knees. He is also chairman of the Department of Anthropology at Harvard, and I make arrangements to meet with him in the departmental office in Cambridge, Massachusetts. I'm hoping that Kleinman will somehow make some sense of this whole issue and miraculously tie the threads of mental illness together in a neat package.

The Harvard campus is surprisingly urban. Traffic whizzes by along a maze of very old streets that surely were not meant to accommodate so many motorized vehicles in need of so few available parking spaces. Students, looking very smart and earnest, make their way from class to class, and there is a smattering of foreign languages in the air. Eventually, I find William James Hall, a towering modern building on the north side of campus that houses the psychology department and the social anthropologists.

After pleasantries, Kleinman invites me into his office, a room jam-packed with books, many of which he wrote. From the first hello I can see that he's an intense man, seemingly always in a hurry, but I soon learn that his intensity comes from the optimistic notion that the world could be a better place. And so we sit, high above the Harvard campus, and talk about how and why people suffer, and what role anthropologists, of all people, might play in helping them.

KLEINMAN IS AN EXPERT IN THE psychiatric disorder called neurasthenia, a condition that was a common diagnosis in Western culture at the turn of the twentieth century. According to Kleinman, this particular diagnosis was a grab bag of symptoms such as fatigue and depression diagnostically bundled as nerves or nervous exhaustion. Neurasthenia was the most fashionable diagnosis of the time; these days we have replaced the term with stress or depression—the name has changed, but the symptoms and the moods are the same.[1]

Although neurasthenia is no longer a valid diagnosis in the West, it is alive and raging in China under the name *shenjing shuairuo*. In his books and articles Kleinman tells of many case studies of patients coming into Chinese clinics to complain of headaches, body aches, tiredness, and such. The feelings are similar to the symptoms of depression in Western culture, but as Kleinman explains, "Many depressed Chinese people do not report feeling sad, but rather express boredom, discomfort, feelings of inner pressure, and symptoms of pain, dizziness, and fatigue."[2] Their complaints, the diagnosis, and the treatment are a more socially acceptable avenue for an individual to express personal and social suffering in a society that disapproves of mental illness. "Mental illness in Chinese culture carries strong stigma," he points out. So the patient receives a diagnosis that doesn't exist in Western medicine, and the Chinese doctors propose various treatments. The diagnosis of neurasthenia in China is just as valid as a diagnosis of depression in the West—both are considered disease states—and yet each culture has different moral values for the symptoms and different treatment options.

What is missing from both diagnoses, Kleinman feels, is an understanding of how the human mind is entwined with the whole experience of being an individual embedded in a family, a culture, and a society. Mind and body cannot be separated, and both are affected by everything else in life. He believes quite strongly that all human experiences are social, and that to understand someone's mental state you must know everything about that person's relationships. But how to do that?

Surprisingly Kleinman thinks that what is absent from health care practice is anthropology—the traditional anthropological techniques of being a careful observer and a good listener, and interested in the details of human relationships. After decades of observing thousands of psychiatric interviews in both Western and Chinese cultures he believes that patients are missing what they need most—compassionate understanding not just of their symptoms but also of their lives.[3] According to Kleinman, mental health is so intimately entwined with everyday experience that it is impossible to treat many symptoms individually because the symptoms are the result of experience, and not really physiological in the classic medical sense. For example, many people, especially poor and powerless people, have no way to express their anger and humiliation and so they "somaticize" their pain: The body becomes the vessel of expression, the mode of releasing anger and frustration, and a way to interact with the world.

Psychotherapists can get a handle on the situation after lots of talk, but anthropologists, according to Kleinman, are better trained to move beyond the psychologically personal to the broader impact of culture and society, which he feels can be just as influential and destructive. "Suffering gives a sense of what are some of the deeper and more significant aspects of social experience," he tells me.

How we are socialized, how cultures and economies change, how we internalize fear and loss according to cultural norms—all have a huge impact on individual mental health. For Kleinman any illness has moral symbolism; every experience is constructed from each person's view of the world and where they fit in. When physicians and mental health workers ignore those very personal aspects of illness, they turn a blind eye to how deeply we are all affected by forces other than viruses, genes, and biochemistry. Sickness and wellness, according to Kleinman, are states of being, states of social interaction, and a mixture of the public and the private. "What we take a symptom to be is a cultural matter, as is the assumption that a symptom mirrors a single deficit in physiological process. That assumption is not only cultural but naive," he explains.[4] The situation is acute in Western culture, Kleinman believes, because this culture has moved swiftly to focus on biological (and most often genetic) explanations for everything from cancer to depression.[5]

Kleinman prefers to think about disease and illness as connected phenomena that might require different treatments. For example, the disease of depression might be helped by antidepressants, but curing the illness is a much more involved process that sometimes means a change of life. "Depression, after all, can be a disease, a symptom, or a normal feeling," he writes.[6] In other words, just calling it a disease is not enough. Kleinman knows that the medical experience itself, including psychiatric diagnosis and psychotherapy, shapes the very course and outcome of a condition. So he is working toward training physicians and others to slow down, listen, and ask more questions. In other words, he wants to anthropologize medicine.

Kleinman has been involved in mental health issues for decades—surely a rather depressing subject—yet he doesn't seem to be the least bit burned out, jaded, or even discouraged. He thinks that everyone, even anthropologists, can make a difference; where the rest of us see sadness and destruction, he sees hope. "I am extremely excited," he says to me several times over the course of our chat, and his enthusiasm is contagious.

I had expected Kleinman to wrap up the anthropology of mental illness for me, but I hadn't expected him to tie a bright bow around it.

Yet our talk did lift my mood. Alternative models of mental illness underscore that both biology in its broadest sense and culture also in its broadest sense are responsible for our mental health and our mental illness. That means there are endless opportunities for improving human mental health.

WESTERN CULTURE, AGAIN

Western culture has, I believe, bought completely into the medical model of mental illness. Is this model wrong? Is it bad? Certainly not. In fact, Western culture can breathe a sigh of relief on many fronts. We are now living in an era in which it is acceptable to talk publicly about mental disorders and negative moods without shame. Depression, anxiety, manic-depression, obsessive-compulsive disorder—these words are now accepted in the Western vernacular.

There is also much hope for treating and even curing many mental illnesses. Psychotropic medications have brought a sense of reality to schizophrenics, saved many others from suicide, and brought hope to those in despair. We have, in Western culture, found our own type of shamans and potions and magic to deal with human unhappiness. Does this make the Western medical model the best model for understanding and treating mental disorders? So far there are no good comparative data showing that the Western medical model is significantly better at helping mental illness than any other cultural model. So we need to be skeptical about claims that the medical model works better than all other models and should therefore be the only model used across the globe.

Why exactly should we be skeptical?

First of all, Western medicine is a science that thrives on categorization, and mental illness as we know it is parceled into disease names according to specific symptoms. Of course, there is nothing intrinsically wrong with putting labels on human conditions. Labels provide a common linguistic ground for talking about people, behaviors, and moods. Labels such as "depression" and "anxiety," for example, give physicians and therapists a familiar base from which to initiate treatment, and they impose order on the world of the patient-client interaction. These labels are as necessary in Western culture as the words "witchcraft" and "sorcery" are in other cultures to explain mental illness. In both cases the labels shift blame away from the victim and bring social legitimacy. Calling a human behavior or mental mood a disease also brings help. For example, only diseases receive health care funding, and with that funding lives are salvaged; so the label is an expedient means to an end.

The downside of labeling human behavior, especially negative mentality, as various diseases is that such categorization really doesn't work—mental illnesses are diseases that are not strongly genetic, not viral, and not bacterial, as far as we know. Yet disease labels direct who gets what treatment and for how long. Use one label—say, "psychosis"—in Western culture and treatment moves down one path; with another label—for example, "neurosis"—there might be no treatment at all. And while labels allow relief from shame—"I have a disease called depression. It's not my fault."—they also obliterate the social content of a mental condition. It's easy to affix a label, harder to face up to the fact that we, as a species, are not very good at curing human misery. Labels just don't say enough, or they don't fit well; labels also become unglued, or ripped, or the words begin to fade into each other when faced with a very real person with very real problems coping or fitting in.

Embedded in the move toward a biomedical model is also the mistaken idea that biology is both the cause and cure of mental illness. It is true that research in genetics is providing some very significant clues about mental health, but it's way too early to really know which genes do what. More important, no one knows what the impact of such information might have on a culture. The Azande belief in witchcraft has infiltrated their culture, and in the same way Western culture is at risk of assuming genes are the answer to everything and that we are just as powerless over our genes as the Azande are over witches. Instead, we might combine the research on monkey behavior with the contributions of genetics research and gain powerful tools for shaping lives. Those with a genetic predisposition toward depression or anxiety could be helped with better parenting and family lives. Those with good genes but bad experiences might be buffered in other ways. It will be a challenge, however, to see if Western culture is willing to embrace an apparently complex notion—the interaction of biology and experience—as the root of much human unhappiness.

Embracing the medical model has also pushed the culture of mental illness into the drugstore, and there is both profound help and grave danger inside those doors. We are currently awash in psychoactive drugs. The exact number of prescriptions written or pills taken is impossible to find for the United States because health care is in the private sector. Psychiatrist Randolph Nesse has looked hard for the statistics, and he discovered that prescriptions for mental disorders in the United States have increased from 131 million in 1988 to 233 million in 1998, and that about 20 million people in this country now take antidepressants.[7] It's easier to track the consumption of medications in countries that have nationalized health care. In Canada, for example, prescriptions for antidepressants increased 64 percent between 1996 and 2000;[8] in Great Britain physicians wrote 8 million more prescriptions for drugs to relieve stress, anxiety, and depression in 2003 than in 1998, a mere five years.[9] These startling increases might actually be good; it may be that more people are being identified and treated. In fact, the use of medication to help people get out of a depression, or to calm psychosis, or to relieve anxiety is a godsend, as anyone who has used the drugs can tell you.

The problem is that we have bought into a biochemical model that is not yet proven, and in our wholesale acceptance we sidestep the social and political context of mental health. For example, there is a strong positive association between poverty and depression. There is also an association between anxiety and cultural change—people who are isolated from their community are miserable. Such social ills are at the root of human misery, and while taking a pill to cope is helpful, it's not the answer. As Javier Escobar has pointed out, the social not only defines how a mental illness is expressed, it also guides the prognosis. Schizophrenics seem to do better in less developed nations because, some researchers suspect, they are not shunned but accepted, not isolated but integrated—and they are not medicated.[10]

There are other clues that should intrigue us and compel us to ask appropriate social questions that move beyond the medical model. For example, we accept as fact the notion that more women than men present with depression. We suspect that women are genetically more vulnerable, or biologically less able to cope with stress—which are culturally torqued assumptions. But the gender difference in depression is found primarily in the United States.[11] Instead of assuming women are biologically at high risk, we might ask why women in the United States in particular are more prone to depression.

We should also be skeptical of pharmaceuticals because they haven't been proven using the very same medical model that assumes their efficacy. Since 1962, the U.S. Food and Drug Administration has required that any new treatments or drugs must be safe and effective. The most common and most well-respected way to prove efficacy is to conduct a double-blind clinical trial, in which those with the condition are given either the real drug or a placebo, and both the patients and the researchers are "blind" to which group received the real thing. In that way, hopeful thinking can't mask any true effect. But people who are on placebos in such trials often feel better, or even show physiological changes that suggest they must have taken the new drug. Researchers have come to expect that about 25 to 30 percent of those on the fake drug will improve thanks to the placebo effect. Just participating in a clinical trial—that is, having hope in some benefit from participating in the trial—often leads to the relief of symptoms. There are many reasons a person might feel better; having hope is one reason, and it can change brain biochemistry independent of the drug.

The idea of a placebo effect, it turns out, is especially important when testing psychoactive drugs. Clinical trials of antidepressants have shown over and over that they aren't any more effective and don't do much more than sugar pills.[12] In one analysis of 19 placebo-controlled clinical trials of antidepressants, Irving Kirsch and Guy Sapirstein found that only 25 percent of the lift in mood was due to the drug, while 50 percent was due to the placebo effect.[13] In other words, when taking either the placebo or the antidepressant, people felt less depressed. And recent research is showing that the placebo effect makes for very real changes in brain biochemistry.[14]

Why should it matter that the placebo effect is so strong for antidepressant medication? The problem is that we now have a false sense that psychotropic medication can fix mental illness, and that assumption reinforces the medical model without any real scientific basis. Implicit in this acceptance is the motivation of drug companies to make huge profits on drugs that have yet to be proven effective[15] and that might even be harmful. In 2005, for example, distraught parents discovered that the pharmaceutical company GlaxoSmithKline had suppressed results of clinical trials on adolescent use of the antidepressant Paxil. It turns out that a drug that was supposed to help their children was often putting them in grave danger; and it may be that all antidepressants carry a high risk of suicide for teens and children.[16] These concerns follow on the heels of questions about the use and misuse of antipsychotic drugs, which are now prescribed for other mental or behavioral problems.[17]

Contrary to the culturally accepted notion that drugs are the best, most reasonable treatment for mental illness, the truth is no one knows how they work or if they really work. More troubling, no one knows what the consequences will be for a culture that has relied on pharmaceuticals to treat mental conditions that are not solely biological but cultural as well. Yet the populace seems to have no trouble downing pills for which there is sketchy research. Such wholesale acceptance makes it more difficult to weed out which clients might best benefit from drugs and which might do better with talk therapy or perhaps a change of life.

The Western medical model of mental illness is dangerous because it is arrogant. We have the best medical facilities, the most sophisticated technologies, and the most comfortable life. Surely we should also be the happiest people on Earth. Yet with all these advantages, it appears that mental illness, especially depression, is ubiquitous in Western culture. By unquestioningly accepting our own belief system, we run the risk of not learning from other cultures that might, in fact, be coping much better. We have to acknowledge that although the Western medical model of mental illness—the disease model— works in some situations and for some people, it's not the only model that has value.

CROSSING CULTURES

Humans used to live in small isolated groups that developed their own brand of culture. But these days the world population is so huge that we can't help but bump into each other, culturally speaking. Many people believe that globalization is the latest stage of human interaction, that every culture on Earth is now economically entwined and we share one large culture—that is, Western culture. Given that perspective, it's easy to assume that all the influence goes in one direction from the economically successful West toward the more economically disadvantaged countries. Money is power, and so the West should, some think, swallow other cultures whole.

The truth is that the influence can go in both directions. Javier Escobar wasn't kidding when he told me that we have a lot to learn from other countries. For example, a recent article in *The New York Times* reported that depression is very rare in Iceland.[18] Iceland is a country with long winters and very short summers, and we might expect Icelanders to be wallowing in depression. But they are not. What makes the story interesting is the way the article is pitched to American readers. Surely, the story claims, scores of Icelanders must be depressed (like us); they just aren't telling the truth. Instead of ridiculing Icelanders, we might dispatch a planeload of psychologists to find out why Icelanders are happier than we are and then import their strategies. In another recent article from the *The New York Times,* Juan Forero reports on the growing teen suicide rate among Amazonian Indians in Colombia.[19] The hook of the article is that indigenous people have the same mental ills as we do in modern Western culture, that teens will be teens. But as the reporter makes clear, these groups are under tremendous social and cultural pressure, and the rash of teen suicides is another example of painful unhappiness for individuals experiencing cultural transition. The lesson to be learned is not that indigenous Colombian Indians are not able to cope with modern life, but that everyone loses faith—and sometimes their mind—when life becomes totally unpredictable, unfamiliar, and without hope.

What other cultures and other models of mental illness—the evolutionary model, the nutritional model, the idea that genes and environment interact to form our attitudes, or the idea that witches are to blame—teach us is that there are all sorts of ways to conceive of human behavior, and all have validity for treatment. These models show that the human mind is universally known to be quite fragile. The mind apparently needs social connections, a stable culture, and a psychological way to cope with life's uncertainties. That stability might be provided by a belief in witchcraft, or psychoanalysis, or psychotropic drugs; each has its advantages and disadvantages. And in every model, there is a means to both lay blame and dispense treatment.

The panoply of models and belief systems to explain human mental states is most of all, I think, a testament to hope. Mental illness is human misery; the opposite of misery is not happiness, but hope. If we listen to the evolutionists and the monkeys, if we eat right, and if we pay attention to cross-cultural psychiatrists, we might just spread that hope across the globe.

NOTES

1 THE ROOTED SORROW

1 (Narrow, Rae et al. 2002).
2 (WHO 2004).
3 (Murthy, Bertolote et al. 2001).
4 (American Psychiatric Association 1994).
5 (Porter 1996).
6 (National Institute on Alcohol Abuse and Alcoholism 2004).
7 (National Institute on Drug Abuse 2004).
8 (Porter 1996; Shorter 1997, 2003).
9 (Porter 1996).
10 (U.S. Department of Health and Human Services 1999).
11 (Porter 1996).
12 (Weatherall 1996).
13 (Gelman 1999).
14 (Shorter 1997).
15 (Nemeroff 1998).
16 (Healy 1999; Greenberg 2003; Nestler 1998; U.S. Department of Health and Human Services 1999; Whitaker 2001; Kirsch and Sapirstein 2003).

17 (Porter 2002).

18 (Healy 1999).

19 With the growing number of children and adolescents taking various pre-
    scribed mood-altering drugs, and the acceptance of such medications for
    growing minds, this culture is also at a crisis point in terms of guiding chil-
    dren. Should children be medicated to make them focused, be more social,
    happier, and more "normal"? Do we medicate the bully and the wallflower?
    If we do, and the children are happier, isn't the life experience of each child
    changed forever? Is that good or bad?

20 (Porter 1996).

21 (Porter 1996).

22 (U.S. Department of Health and Human Services 1999).

23 (Chambless, Sanderson et al. 1996).

24 (Porter 1996).

25 (Porter 1996).

26 (SAMSHA 2003). These data are from 2001. Hawaii spends almost as much
    as New York state ($175.21 per person) on mental health, but the next-
    ranked state, Pennsylvania, spends $151.98. West Virginia spends the least at
    $25.52 per capita.

27 (Jablensky, Sartorius et al. 1992). This finding, however, has been disputed
    (Edgerton and Cohen 1994).

28 (Minksy, Vega et al. 2003).

29 (Burnam, Holman et al. 1987; Burnam, Hough et al. 1987; Vega, Kolody et
    al. 1998; Escobar, Gara et al. in press).

30 (Escobar 1998).

31 (Kendell 1975).

32 (Payer 1988).

2 THE EVOLUTION OF THE MIND

1 (Deacon 1992).

2 (Tattersall 1993).

3 (Cheney, Seyfarth et al. 1986).

4 (Eisenberg 1981).

5 (Deacon 1992).

6 (Gilbert 1998a,b; Fábrega 2002).

7 (Nesse 1990; Gilbert 1998a,b).

8 (Darwin 1872; Crow 1995).

9   (McGuire and Troisi 1998).

10   (Lewis 1936).

11   (Nesse 1984).

12   (Barkow, Cosmides et al. 1992; Pinker 1997).

13   (McGuire and Troisi 1998).

14   (Gilbert 1998a), p. 351.

15   (Nesse 1991).

16   (Nesse 1991), p. 33.

17   (Nesse and Lloyd 1992).

18   (Nesse and Williams 1996; Gilbert 1998a,b).

19   (Nesse and Lloyd 1992).

20   (Nesse 1998).

21   (Byrne 1995).

22   (Shorter 1997).

23   (American Psychiatric Association 1994; U.S. Department of Health and Human Services 1999; Javitt and Coyle 2004).

24   In this dynamic field, the best reference is the Web site *schizophrenia.com*, which tracks all the latest international research results.

25   (American Psychiatric Association 1994; McGuire and Troisi 1998).

26   (McGuffin, Owen et al. 1995; Portin and Alanen 1997).

27   (Wilson 1993, 1998; McGuire and Troisi 1998).

28   (Crow 1995).

29   (Allen and Sarich 1988; Crow 1995).

30   (Stevens and Price 1996).

31   (McGuire and Troisi 1998).

32   (McGuire and Troisi 1998), p. 212.

33   (Javitt and Coyle 2004).

34   (Murthy, Bertolote et al. 2001).

35   (Nesse 1984; Stevens and Price 1996).

36   (Cannon 1929).

37   (Stevens and Price 1996), p. 98.

38   (Marks and Nesse 1994; Nesse 1999b).

39   (Nesse 1984).

40   (Stevens and Price 1996). See also Chapter 3.

41   (Nesse 1990), p. 282.

42   (Marks and Nesse 1994).

43   (Marks 2002), p. 254.

44 (Marks 2002).
45 (Group 1992).
46 (Kessler 2003).
47 (Nesse 2000).
48 (Hagen 2003).
49 (Bowlby 1969).
50 (Hagen 2003).
51 (Price 1967).
52 (Price, Sloman et al. 1994; Price 1998).
53 (Gilbert and Allen 1998).
54 (Nesse 1999a).

3 THE MINDS OF MONKEYS

1 (Jolly 1986; Smuts, Cheney et al. 1987; Arcadi in prep.).
2 (Goodall 1986).
3 (Marks 2002).
4 The best description of Harlow's life and the impact of his experiments can be found in Debora Blum's fine book, *Love at Goon Park* (2002).
5 (Bowlby 1969).
6 (Harlow, Dodsworth et al. 1965).
7 (Harlow, Plubell et al. 1973; Harlow and Suomi 1974).
8 (Suomi, Eisele et al. 1975).
9 (Harlow and Suomi 1971a).
10 (Harlow and Novak 1973).
11 (Harlow, Dodsworth et al. 1965), p. 96.
12 (Blum 2002 #92).
13 (Blum 2002 #92).
14 (Suomi, Harlow et al. 1974).
15 (Harlow and Suomi 1971b).
16 (Small 1998).
17 (Small 1993).
18 (Cheney, Seyfarth et al. 1986).
19 (Small 1993).
20 (Hamilton 1963).
21 (Small 1993).

22 (Smuts, Cheney et al. 1987).

23 (Byrne and Whiten 1988).

24 (deWaal 1996).

25 See also the work on genes that control for the expression of the enzyme monoamine oxidase A (MAOA), which is also implicated in aggression in humans and monkeys (Barr, Becker et al. 2003).

26 (Lesch, Bengel et al. 1996).

27 (Bennett, Lesch et al. 2002).

28 (Caspi, Sugden et al. 2003).

29 (Heinz, Jones et al. 2003; Barr, Newman et al. 2004).

30 (Caspi, Sugden et al. 2003).

31 (Mehlman, Higely et al. 1995; Higely, Mehlman et al. 1996).

32 (Mehlman, Higely et al. 1994).

4 THE HAPPY FAT

1   (Klerman and Weissman 1989; Group 1992).

2   (Salem 1999).

3   (de la Presa Owens and Innis 1999).

4   (Ikemoto, Kobayashi et al. 1997; Kim, Akbar et al. 2000; Yoshida, Miyazaki et al. 2001).

5   (Tsukada, Kakiuchi et al. 2000).

6   (Hibbeln, Umhau et al. 1997).

7   (Hibbeln, Umhau et al. 1998).

8   (Hibbeln and Salem 1995).

9   (Hibbeln 1998).

10 (Cott and Hibbeln 2001; Magnusson, Axelsson et al. 2000).

11 (Hibbeln, Enstrom et al. 2000; Tanskanen and Hibbeln 2001; Hibbeln 2002; Noaghiul and Hibbeln 2003).

12 (Buydens-Branchey, Branchy et al. 2003).

13 (Hibbeln in press).

14 (Stoll, Severus et al. 1999).

15 (Nemets, Stahl et al. 2002).

16 (Peet, Brind et al. 2001).

17 (Mellor, Laugharne et al. 1996).

18 (Puri, Steiner et al. 1998).

19 (Morris, Evans et al. 2003).

20 (Helland, Smith et al. 2003).
21 (Carlson, Cooke et al. 1992; Agostoni, Riva et al. 1995; Carlson, Ford et al. 1996; Birch, Garfield et al. 2000).
22 (Iribarren, Markowitz et al. 2004).
23 (Crawford, Bloom et al. 1999).
24 (Eaton 1992; Broadhurst, Cunnane et al. 1998; Broadhurst, Wang et al. 2002).
25 (Cordain, Watkins et al. 2001; Cordain, Watkins et al. 2002).
26 (Gibbons 2002).
27 (Walter, Buffler et al. 2000).
28 (Brooks, Helgren et al. 1995; Gibbons 2002).
29 (Richards, Pettitt et al. 2001).
30 (Eaton, Eaton et al. 1998).
31 (Cordain, Watkins et al. 2002).
32 (Reis and Hibbeln submitted).

5 STATES OF MIND

1   (Kane 1995).
2   (Konner 1995).
3   (Kroeber and Kluckhohn 1952).
4   (Kessing and Strathern 1998).
5   (Ferraro, Trevethan et al. 1994).
6   (Marks 2002).
7   There are many books about Boas, but the most lucid account of his influence on American culture can be found in Pierpont (2004).
8   (Marks 2002), p. 90.
9   (Marks 2002), p. 91.
10 (Marks 1995).
11 (Goodall 1986).
12 (Whiten, Goodall et al. 1999).
13 (McGrew 1992; Small 2000).
14 (van Schaik, Ancrenaz et al. 2003).
15 (Kawai 1965).
16 (Tattersall 1993).
17 (Tattersall 1993).
18 (Ehrlich and Feldman 2003).

19 (Jolly 1966; Cheney, Seyfarth et al. 1986).
20 (Kaplan, Hill et al. 2000; Leonard 2002; Sanford 2003).
21 Adapted from Gottlieb (2004).
22 (Small 1993).
23 (Kagan 1989; Kagan and Sniderman 1991a,b; Kagan, Sniderman et al. 1992).
24 (Ridley 2000).
25 (Ridley 2003).
26 (Kessing and Strathern 1998), p. 49.
27 (LeVine, Dixon et al. 1994).
28 (Harkness and Super 1997).
29 (Panter-Brick and Smith 2000).
30 (Flinn and England 1995, 1997).
31 (Howard 1984).
32 (Benedict 1934a).
33 (Benedict 1928).
34 (Benedict 1934b), p. 232.
35 (Benedict 1934a).
36 (Benedict 1934a).
37 (Benedict 1934b), p. 254.

6 RUNNING AMOK IN A BRAIN FOG

1 (Gaw 2001).
2 (Gaw 2001).
3 (Levine and Gaw 1995).
4 (Hughes 1985).
5 (Yap 1962; Gaw 2001).
6 For a wonderful and intimate story about the interface between Western medicine and immigrant culture, see Fadimin (1998).
7 (Linde 2002),
8 (Carr and Tan 1976; Gaw and Bernstein 1992; Gaw 2001).
9 (Gimlette 1901; Gaw and Bernstein 1992).
10 (Arboleda-Florez 1985).
11 (Teoh 1972).
12 (Westermeyer 1972; Arboleda-Florez 1985).
13 (Westermeyer 1972; Arboleda-Florez 1985).

14   (Arboleda-Florez 1985).
15   (Mahathir 1970; Arboleda-Florez 1985).
16   (Carr and Tan 1976).
17   (Gaw and Bernstein 1992).
18   (Teoh 1972).
19   (Simons 1996).
20   (Osborne 2001).
21   (Geertz 1968).
22   (Siegel 1986).
23   (Geertz 1968; Siegel 1986).
24   (Simons 1996).
25   (Kenny 1985; Osborne 2001).
26   (Gaw 2001; Osborne 2001).
27   (Simons 1996), p. 247.
28   (Tseng, Kan-Ming et al. 1988; Bernstein and Gaw 1990).
29   (Gaw 2001; Tseng 2001).
30   (Gaw 2001; Tseng 2001).
31   (Jilek and Jilek-Aal 1985).
32   (Bernstein and Gaw 1990).
33   (Tseng, Kan-Ming et al. 1988).
34   (Jilek and Jilek-Aal 1985).
35   (Gaw 2001).
36   (Landy 1985).
37   (Gaw 2001).
38   (Brumberg 1988).
39   (Brumberg 1992).
40   (Brumberg 1988).
41   (Rittenbaugh 1982).
42   (American Psychiatric Association 1994; Gaw 2001).
43   (Low 1985).

7 CURSED AND HAUNTED

1   (Spiro 1967, 1970).
2   (Spiro 1967).
3   (Gaw 2001).

4   (Suwanlert 1976).

5   (Suwanlert 1976), p. 75.

6   (Gaw 2001).

7   (Spiro 1967), p. 229.

8   (Suwanlert 1976).

9   (Behringer 2004).

10 (Gaw 2001).

11 (Ferraro, Trevethan et al. 1994).

12 (Evans-Pritchard 1937), p. 21.

13 (Evans-Pritchard 1937), p. 22.

14 (Evans-Pritchard 1937), p. 21.

15 (Evans-Pritchard 1937), p. 91.

16 (Evans-Pritchard 1937), p. 63.

17 (Evans-Pritchard 1937), p. 64.

18 (Guenther 1999).

19 (Service 1963).

20 (Hutton 2001).

21 (Lewis 1981).

22 (Silverman 1967).

23 (Lee 1979, 1984).

24 (Lee 1984), p. 109.

25 (Lee 1967).

26 (Lee 1967), p. 37.

27 (Bourguignon 1976).

8 A HAPPY ENDING

1   (Kleinman 1988a).

2   (Kleinman 2004).

3   (Kleinman 1995).

4   (Kleinman 1988b), p. 8.

5   (Kleinman 1988b).

6   (Kleinman 1988b), p. 16.

7   (Nesse 2005).

8   (Hemels, Koren et al. 2002).

9   (Wright 2003).

10   Others feel this conclusion is suspect. See Edgerton and Cohen (1994).
11   (McGuire and Troisi 1998).
12   (Greenberg 2003).
13   (Kirsch and Sapirstein 2003).
14   (Mayberg, Silva et al. 2002).
15   (Goode 2002; Koerner 2003).
16   (Carey 2004; Harris 2004).
17   (Goode 2003).
18   (Lyall 2004).
19   (Forero 2004).

REFERENCES

Agostoni, C., E. Riva, et al. (1995). "Docosahexaenoic acid status and developmental quotient of healthy term infants." *Lancet* **346**: 638.

Allen, J. S., and V. M. Sarich (1988). "Schizophrenia in an evolutionary perspective." *Perspectives in Biology and Medicine* **32**: 132-151.

American Psychiatric Association (1994). *Diagnostic and Statistical Manual of Mental Disorders, Fourth Edition: DSM-IV*. Washington, D.C., American Psychiatric Association.

Arboleda-Florez, J. (1985). Amok. *The Culture-Bound Syndromes*. R. C. Simons and C. C. Hughes, eds. Dordrecht, The Netherlands, D. Reidel Publishing Company: 251-264.

Arcadi, A. C. (in prep.). "Nonhuman primate behavioral plasticity and the evolution of 'free will.'" *Current Anthropology*.

Barkow, J. H., L. Cosmides, et al. (1992). *The Adapted Mind*. Oxford, Oxford University Press.

Barr, C. S., M. L. Becker, et al. (2003). "Relationships among CSF monoamine metabolite levels, alcohol sensitivity, and alcohol-related aggression in rhesus macaques." *Aggressive Behavior* **29**: 288-301.

Barr, S. B., T. K. Newman, et al. (2004). "Rearing condition and rh5-HTTLPR interact to influence limbic-hypothalamic-pituitary-adrenal axis response to stress in infant macaques." *Biological Psychiatry* **55**: 733-738.

Behringer, W. (2004). *Witches and Witch-Hunts: A Global History.* New York, Polity Press.

Benedict, R. (1928). "Psychological types in the cultures of the Southwest." *Proceedings of the Twenty-Third International Congress Americanists* **September**: 527-581.

Benedict, R. (1934a). "Anthropology and the abnormal." *Journal of General Psychology* **10**: 59-82.

Benedict, R. (1934b). *Patterns of Culture.* New York, Houghton Mifflin.

Bennett, A. J., K. P. Lesch, et al. (2002). "Early experience and serotonin transporter gene variation interact to influence primate CNS function." *Molecular Psychiatry* **7**: 118-122.

Bernstein, R. L., and A. L. Gaw (1990). "Koro: Proposal classification for DSM-IV." *American Journal of Psychiatry* **147**: 1670-1674.

Birch, E. E., S. Garfield, et al. (2000). "A randomized controlled trial on early dietary supply of long-chain polyunsaturated fatty acids and mental development in term infants." *Developmental Medicine and Child Neurology* **42**: 174-181.

Blum, D. (2002). *Love at Goon Park: Harry Harlow and the Science of Affection.* New York, Perseus.

Bourguignon, E. (1976). Possession and trance in cross-cultural studies of mental health. *Culture-Bound Syndromes, Ethnopsychiatry, and Alternate Therapies.* W. P. Lebra, ed. Honolulu, University Press of Hawaii: 47-55.

Bowlby, J. (1969). *Attachment and Loss.* New York, Basic Books.

Broadhurst, C. L., S. C. Cunnane, et al. (1998). "Rift Valley lake fish and shellfish provided brain-specific nutrition for early *Homo*." *British Journal of Nutrition* **79**: 3-21.

Broadhurst, C. L., Y. Wang, et al. (2002). "Brain-specific lipids from marine, lacustrine, or terrestrial food resources: Potential impact on early African *Homo sapiens*." *Comparative Biochemistry and Physiology* **131** (**B**): 653-673.

Brooks, A., D. M. Helgren, et al. (1995). "Dating and context of three Middle Stone Age sites with bone points in the Upper Semliki Valley, Zaire." *Science* **268**: 548-556.

Brumberg, J. J. (1988). *Fasting Girls: The Emergence of Anorexia Nervosa as a Modern Disease.* Cambridge, Harvard University Press.

Brumberg, J. J. (1992). From psychiatric syndrome to "communicable" disease: The case of anorexia nervosa. *Framing Disease.* C. Rosenberg and J. Golden, eds. New Brunswick, N.J., Rutgers University Press: 134-154.

Burnam, M. A., A. Holman, et al. (1987). "Six-month prevalence of specific psychiatric disorders among Mexican Americans and non-Hispanic whites in Los Angeles." *Archives of General Psychiatry* 44: 682-694.

Burnam, M. A., R. L. Hough, et al. (1987). "Acculturation and lifetime prevalence of psychiatric disorders among Mexican Americans." *Journal of Health, Society and Behavior* 28: 89-102.

Buydens-Branchey, L., M. Branchy, et al. (2003). "Polyunsaturated fatty acid status and relapse vulnerability in cocaine addicts." *Psychiatry Research* 120: 29-35.

Byrne, R. (1995). *The Thinking Ape.* Oxford, Oxford University Press.

Byrne, R. W., and A. Whiten (1988). *Machiavellian Intelligence: Social Expertise and the Evolution of Intelligence in Monkeys, Apes and Humans.* Oxford, Clarendon Press.

Cannon, W. B. (1929). *Bodily Changes in Pain, Hunger, Fear and Rage: Researches into the Functions of Emotional Excitement.* New York, Appleton.

Carey, B. (2004). "Is Prozac better? Is it even different?" *The New York Times* September 21.

Carlson, S. E., R. J. Cooke, et al. (1992). "First year growth of preterm infants fed standard compared to marine oil n-3 supplemental formula." *Lipids* 27: 901-907.

Carlson, S. E., A. J. Ford, et al. (1996). "Visual acuity and fatty acid status of term infants fed human milk and formulas with and without docosahexaenoate and arachidonate from egg yolk lecithin." *Pediatric Research* 39: 882-888.

Carr, J. E., and E. K. Tan (1976). "In search of the true amok: Amok as viewed within the Malay culture." *American Journal of Psychiatry* 133: 1295-1299.

Caspi, A., K. Sugden, et al. (2003). "Influence of life stress on depression: Moderation by a polymorphism in the 5-HTT gene." *Science* 301: 386-389.

Chambless, D. L., W. C. Sanderson, et al. (1996). "An update on empirically validated therapies." *The Clinical Psychologist* 49: 5-18.

Cheney, D. L., R. M. Seyfarth, et al. (1986). "Social intelligence and the evolution of the primate brain." *Science* **243**: 1361-1366.

Cordain, L., B. A. Watkins, et al. (2001). Fatty acid composition and energy density of foods available to African hominids. *Nutrition and Fitness: Metabolic Studies in Health and Disease.* A. P. Simopoulos and K. N. Pavlou, eds. Basel, Switzerland, Karger. **90**: 144-161.

Cordain, L., B. A. Watkins, et al. (2002). "Fatty acids analysis of wild ruminant tissues: Evolutionary implications for reducing diet-related chronic disease." *European Journal of Clinical Nutrition* **56**: 181-191.

Cott, J., and J. R. Hibbeln (2001). "Lack of seasonal affective disorder in Icelanders." *American Journal of Psychiatry* **158**: 328.

Crawford, M. A., M. Bloom, et al. (1999). "Evidence of the unique functions of DHA during the evolution of the modern hominid brain." *American Journal of Clinical Nutrition* 1-19.

Crow, T. J. A. (1995). "A Darwinian approach to the origins of psychosis." *British Journal of Psychiatry* **167**: 12-25.

Darwin, C. (1872). *The Expression of Emotions in Man and Animals.* Chicago, University of Chicago Press.

de la Presa Owens, S., and S. M. Innis (1999). "Docosahexaenoic and arachidonic acid prevent a decrease in dopaminergic and serotoninergic neurotransmitters in frontal cortex caused by a linolenic and alpha-linolenic acid deficient diet in formula-fed piglets." *Journal of Nutrition* **129**: 2088-2093.

Deacon, T. (1992). The human brain. *The Cambridge Encyclopedia of Human Evolution.* S. Jones, R. Martin, and D. Pilbeam, eds. Cambridge, Cambridge University Press: 115-123.

deWaal, F. (1996). *Good Natured: The Origins of Right and Wrong in Humans and Other Animals.* Cambridge, Harvard University Press.

Eaton, S. B. (1992). "Humans, lipids and evolution." *Lipids* **27**: 814-820.

Eaton, S. B., S. B. Eaton, et al. (1998). Dietary intake of long-chain polyunsaturated fatty acids during the Paleolithic. *The Return of w3 Fatty Acids into the Food Supply. I. Land-Based Animal Food Products and Their Health Effects.* A. P. Simopoulos, ed. Basel, Switzerland, Karger. **83**: 12-23.

Edgerton, R. B., and A. Cohen (1994). "Culture and schizophrenia: The DOSMD challenge." *British Journal of Psychiatry* **164**: 222-231.

Ehrlich, P., and M. Feldman (2003). "Genes and culture." *Current Anthropology* **44**: 87-107.

Eisenberg, J. F. (1981). *The Mammalian Radiation.* Chicago, University of Chicago Press.

Escobar, J. I. (1998). "Immigration and mental health: Why are immigrants better off?" *Archives of General Psychiatry* **55**: 781-782.

Escobar, J. I., M. Gara, et al. (in press). "Somatization in primary care." *British Journal of Psychiatry.*

Evans-Pritchard, E. E. (1937). *Witchcraft, Oracles and Magic Among the Azande.* Oxford, Oxford University Press.

Fábrega, H. (2002). *Origin of Psychopathology: The Phylogenetic and Cultural Basis of Mental Illness.* New Brunswick, N.J., Rutgers University Press.

Fadimin, A. (1998). *The Spirit Catches You and You Fall Down.* New York, Farrar, Straus and Giroux.

Ferraro, G., W. R. Trevethan, et al. (1994). *Anthropology: An Applied Perspective.* St. Paul, Minn., West Publishing Company.

Flinn, M. V., and B. G. England (1995). "Childhood stress and family environment." *Current Anthropology* **36**: 854-866.

Flinn, M. V., and B. G. England (1997). "The social economics of childhood glucocorticoid stress response and health." *American Journal of Physical Anthropology* **102**: 33-53.

Forero, J. (2004). "In a land torn by violence, too many troubling deaths." *The New York Times* **November 23**.

Gaw, A. C. (2001). *Cross-Cultural Psychiatry.* Washington, D.C., American Psychiatric Publishing, Inc.

Gaw, A. C., and R. B. Bernstein (1992). "Classification of amok in the DSM-IV." *Hospital and Community Psychiatry* **43**: 789-793.

Geertz, H. (1968). "*Latah* in Java: A theoretical paradox." *Indonesia* **5**: 93-104.

Gelman, S. (1999). *Medicating Schizophrenia.* New Brunswick, N.J., Rutgers University Press.

Gibbons, A. (2002). "Human's head start: New views of brain evolution." *Science* **296**: 835-837.

Gilbert, P. (1998a). "Evolutionary psychopathology: Why isn't the mind designed better than it is?" *British Journal of Medical Psychology* **71**: 353-373.

Gilbert, P. (1998b). "Preface and outline." *British Journal of Medical Psychology* **71**: 351-352.

Gilbert, P., and J. S. Allen (1998). "The role of defeat and entrapment (arrested flight) in depression: An exploration of an evolutionary view." *Psychological Medicine* **28**: 585-598.

Gimlette, J. D. (1901). "Notes on a case of amok." *Journal of Tropical Medicine and Hygiene* **4**: 195-199.

Goodall, J. (1986). *The Chimpanzees of Gombe.* Cambridge, Harvard University Press.

Goode, E. (2002). "Antidepressants lift clouds, but lose 'miracle drug' label." *The New York Times* **June 30**.

Goode, E. (2003). "Leading drugs for psychosis come under new scrutiny." *The New York Times* **May 20**.

Gottlieb, A. (2004). *The Afterlife Is Where We Come From: The Culture of Infancy in West Africa.* Chicago, University of Chicago Press.

Greenberg, G. (2003). "Is it Prozac or placebo?" *Mother Jones* **November/ December**: 77-81.

Group, C.-N. C. (1992). "The changing rate of major depression: Cross-national comparisons." *Journal of the American Medical Association* **268**: 3098-3105.

Guenther, M. (1999). From totemism to shamanism: Hunter-gatherer contributions to world mythology and spirituality. *The Cambridge Encyclopedia of Hunters and Gatherers.* R. B. Lee and R. Daly, eds. Cambridge, Cambridge University Press: 426-433.

Hagen, E. H. (2003). The bargaining model of depression. *Genetic and Cultural Evolution of Cooperation.* P. Hammerstein, ed. Boston, MIT Press: 95-123.

Hamilton, W. D. (1963). "The evolution of altruistic behavior." *American Naturalist* **97**: 354-356.

Harkness, S., and C. M. Super (1997). "Why African children are so hard to test." *Annals of the New York Academy of Sciences* **285**: 326-331.

Harlow, H. F., and M. A. Novak (1973). "Psychopathological perspectives." *Perspectives in Biology and Medicine* **Spring**: 461-478.

Harlow, H. F., and S. J. Suomi (1971a). "Production of depressive behaviors in young monkeys." *Journal of Autism and Childhood Schizophrenia* **1**: 246-255.

Harlow, H. F., and S. J. Suomi (1971b). "Social recovery by isolation-reared monkeys." *Proceedings of the National Academy of Sciences* **68**: 1534-1538.

Harlow, H. F., and S. J. Suomi (1974). "Induced depression in monkeys." *Behavioral Biology* **12**: 273-296.

Harlow, H. R., R. O. Dodsworth, et al. (1965). "Total isolation in monkeys." *Proceedings of the National Academy of Sciences* **54**: 90-97.

Harlow, H. F., P. E. Plubell, et al. (1973). "Introduction of psychological death in rhesus monkeys." *Journal of Autism and Childhood Schizophrenia* **3**: 299-307.

Harris, G. (2004). "F.D.A. seeks suicide caution for anti-depressants." *The New York Times* **September 24**.

Healy, D. (1999). *The Antidepressant Era*. Cambridge, Harvard University Press.

Heinz, A., D. W. Jones, et al. (2003). "Serotonin transporter availability correlates with alcohol intake in non-human primates." *Molecular Psychiatry* **8**: 231-234.

Helland, I. B., L. Smith, et al. (2003). "Maternal supplementation with very-long-chain n-3 fatty acids during pregnancy and lactation segments children's IQ at 4 years of age." *Pediatrics* **111**: 39-44.

Hemels, M. E., G. Koren, et al. (2002). "Increased use of antidepressants in Canada." *Annual Pharmacotherapy* **36**: 1653.

Hibbeln, J. R. (1998). "Fish consumption and major depression." *Lancet* **351**: 1213.

Hibbeln, J. R. (2002). "Seafood consumption, the DHA content of mother's milk and prevalence of rate of postpartum depression: A cross-national, ecological analysis." *Journal of Affective Disorders* **69**: 15-29.

Hibbeln, J. R. (in press). "Omega-3 status and CSF-CRH in perpetrators of domestic violence." *Biological Psychiatry*.

Hibbeln, J. R., and N. Salem (1995). "Dietary polyunsaturated fatty acids and depression: When cholesterol does not satisfy." *American Journal of Clinical Nutrition* **62**: 1-9.

Hibbeln, J. R., J. C. Umhau, et al. (1997). Do plasma polyunsaturates predict hostility and depression? *Nutrition and Fitness: Metabolic and Behavioral Aspects in Health and Disease*. A. Simopoulos and K. N. Pavlou, eds. Basel, Switzerland, Karger: 175-186.

Hibbeln, J. R., J. C. Umhau, et al. (1998). "A replication study of violent and nonviolent subjects: Cerebrospinal fluid metabolites of serotonin and dopamine are predicted by plasma essential fatty acids." *Biological Psychiatry* **44**: 243-249.

Hibbeln, J. R., G. Enstrom, et al. (2000). *Suicide Attempters and PUFAs: Lower Plasma Eicosapentaenoic Acid Alone Predicts Greater Psychopathology*. 4th Congress of the International Society for the Study of Fatty Acids and Lipids, Tsjkuba, Japan.

Higely, J. D., P. T. Mehlman, et al. (1996). "Excessive mortality in young free-ranging male nonhuman primates with low cerebrospinal fluid 5-hydroxyindoleacetic acid concentrations." *Archives of General Psychiatry* **53**: 537-543.

Howard, J. (1984). *Margaret Mead: A Life*. New York, Simon and Schuster.

Hughes, C. C. (1985). Culture bound or construct bound? *The Culture-Bound Syndromes*. R. C. Simons and C. C. Hughes, eds. Dordrecht, The Netherlands, D. Reidel Publishing Company: 3-24.

Hutton, R. (2001). *Shamans: Siberian Spirituality and the Western Imagination*. London, Hambledon and London.

Ikemoto, A., T. Kobayashi, et al. (1997). "Membrane fatty acid modifications of PC12 cells by arachidonate or docosahexaenoate affect neurite growth but not norepinephrine release." *Neurochemical Research* **22**: 671-678.

Iribarren, C., J. H. Markowitz, et al. (2004). "Dietary intake of n-3, n-6 fatty acids and fish: Relationship with hostility in young adults—the CARDIA study." *European Journal of Clinical Nutrition* **58**: 24-31.

Jablensky, A., N. Sartorius, et al. (1992). "Schizophrenia: Manifestations, incidence and course in different cultures." *World Health Organization* **Supplement #20**.

Javitt, D. C., and J. T. Coyle (2004). "Decoding schizophrenia." *Scientific American* **January**: 48-55.

Jilek, W. G., and L. Jilek-Aal (1985). "The metamorphosis of 'culture-bound' syndromes." *Social Science Medicine* **21**: 205-210.

Jolly, A. (1966). "Lemur social behavior and primate intelligence." *Science* **153**: 501-506.

Jolly, A. (1986). *The Evolution of Primate Behavior*. New York, Macmillan.

Kagan, J. (1989). "Temperamental contribution to social behavior." *American Psychologist* **44**: 668-674.

Kagan, J., and N. Sniderman (1991a). "Infant predictors of inhibited and uninhibited profiles." *Psychological Science* **2**: 40-44.

Kagan, J., and N. Sniderman (1991b). "Temperamental factors in human development." *American Psychologist* **46**: 856-862.

Kagan, J., N. Sniderman, et al. (1992). "Initial reactions to unfamiliarity." *Current Directions in Psychological Science* **1**: 171-1874.

Kane, J. (1995). *Savages*. New York, Knopf.

Kaplan, H., K. Hill, et al. (2000). "A theory of human life history evolution: Diet, intelligence, and longevity." *Evolutionary Anthropology* **9**: 156-185.

Kawai, M. (1965). "Newly acquired pre-cultural behavior of the natural troop of Japanese monkeys on Koshima Islet." *Primates* **1**: 1-30.

Kendell, R. E. (1975). "Psychiatric diagnosis in Britain and the United States." *British Journal of Psychiatry* **Special Publication 9**: 533-561.

Kenny, M. G. (1985). Paradox lost: The *latah* problem revisited. *The Culture-bound Syndromes: Folk Illness of Psychiatric and Anthropological Interest*. R. C. Simons and C. C. Hughes, eds. Dordrecht, The Netherlands, D. Reidel Publishing Company: 63-76.

Kessing, R. M., and A. J. Strathern (1998). *Cultural Anthropology: A Contemporary Perspective*. New York, Harcourt Brace.

Kessler, R. C. (2003). "Epidemiology of women and depression." *Journal of Affective Disorders* **74**: 5-13.

Kim, H. Y., M. Akbar, et al. (2000). "Inhibition of neuronal apoptosis by docosahexaenoic acid (22:6-n-3)." *Journal of Biochemistry* **275**: 35215-35223.

Kirsch, I., and G. Sapirstein (2003). "Listening to Prozac but hearing placebo: A meta-analysis of antidepressant medication." *Prevention and Treatment* **1**: 1-14.

Kleinman, A. (1988a). *The Illness Narratives: Suffering, Healing, and the Human Condition*. New York, Basic Books.

Kleinman, A. (1988b). *Rethinking Psychiatry: From Cultural Category to Personal Experience*. New York, The Free Press.

Kleinman, A. (1995). *Writing at the Margin: Discourse Between Anthropology and Medicine*. Berkeley, University of California Press.

Kleinman, A. (2004). "Culture and depression." *New England Journal of Medicine* **351**: 951-953.

Klerman, G. L., and M. M. Weissman (1989). "Increasing rates of depression." *Journal of the American Medical Association* **261**: 2229-2235.

Koerner, B. I. (2003). Disorders made to order. *Best American Science Writing, 2003*. O. Sacks, ed. New York, HarperCollins: 194-203.

Konner, M. J. (1995). Anthropology and psychiatry. *Comprehensive Textbook of Psychiatry.* H. I. Kaplan and J. Benjamin, eds. Baltimore, Williams & Wilkins. **1**: 337-356.

Kroeber, A. L., and C. Kluckhohn (1952). "Culture: A critical review of concepts and definitions." *Peabody Museum Papers* **47**.

Landy, D. (1985). "*Pibloktoq* (hysteria) and Inuit nutrition: Possible implication of hypervitaminosis A." *Social Science Medicine* **21**: 173-185.

Lee, R. B. (1967). "Trance cure of the !Kung Bushmen." *Natural History* **November**: 31-37.

Lee, R. B. (1979). *The !Kung San: Men, Women, and Work in a Foraging Society.* Cambridge, Cambridge University Press.

Lee, R. B. (1984). *The Dobe !Kung.* New York, Holt, Rinehart and Winston.

Leonard, W. R. (2002). "Food for thought." *Scientific American* **December**: 75-83.

Lesch, K., D. Bengel, et al. (1996). "Association of anxiety-related traits with a polymorphism in the serotonin transporter gene regulatory region." *Science* **274**: 1527-1531.

LeVine, R. A., S. Dixon, et al. (1994). *Child Care and Culture: Lessons from Africa.* Cambridge, Cambridge University Press.

Levine, R. E., and A. C. Gaw (1995). "Culture-bound syndromes." *Psychiatric Clinics of North America* **18**: 523-536.

Lewis, A. J. (1936). "A clinical survey of depressive states." *Journal of Mental Science* **80**: 273-378.

Lewis, I. M. (1981). "What is a shaman?" *Folk* **23**: 25-35.

Linde, P. R. (2002). *Of Spirits and Madness: An American Psychiatrist in Africa.* New York, McGraw-Hill.

Low, S. M. (1985). "Culturally interpreted symptoms or culture-bound syndromes: A cross-cultural review of nerves." *Social Science Medicine* **21**: 187-196.

Lyall, S. (2004). "If no Icelanders admit to feeling blue, are they?" *The New York Times* **September 2**.

Magnusson, A., J. Axelsson, et al. (2000). "Lack of seasonal mood change in the Icelandic population: Results of a cross-sectional study." *American Journal of Psychiatry* **157**: 243-248.

Mahathir, M. (1970). *The Malay Dilemma.* Singapore, Asia Pacific Press.

Marks, I. M., and R. M. Nesse (1994). "Fear and fitness: An evolutionary analysis of anxiety disorders." *Ethology and Sociobiology* **15**: 247-261.

Marks, J. (1995). *Human Biodiversity: Genes, Race and History.* New York, Aldine de Gruyter.

Marks, J. (2002). *What It Means to Be 98% Chimpanzee.* Berkeley, University of California Press.

Mayberg, H., J. A. Silva, et al. (2002). "The functional neuroanatomy of the placebo effect." *American Journal of Psychiatry* 159: 728-737.

McGrew, W. (1992). *Chimpanzee Material Culture.* Cambridge, Cambridge University Press.

McGuffin, P., M. J. Owen, et al. (1995). "Genetic basis of schizophrenia." *Lancet* 346: 678-682.

McGuire, M., and A. Troisi (1998). *Darwinian Psychiatry.* Oxford, Oxford University Press.

Mehlman, P. T., J. D. Higely, et al. (1994). "Low CSF 5-HIAA concentrations, severe aggression and impaired impulse control in nonhuman primates." *American Journal of Psychiatry* 151: 1485-1491.

Mehlman, P. T., J. D. Higely, et al. (1995). "Correlations of CSF 5-HIAA concentrations with sociality and the timing of emmigration in free-ranging primates." *American Journal of Psychiatry* 152: 907-913.

Mellor, J. E., J. D. Laugharne, et al. (1996). "Omega-3 fatty acid supplementation in schizophrenic patients." *Human Psychopharmacology; Clinical and Experimental* 11: 39-46.

Minksy, S., W. A. Vega, et al. (2003). "Diagnostic patterns in Latino, African American, and European American psychiatric patients." *Archives of General Psychiatry* 60: 637-644.

Morris, M. C., D. A. Evans, et al. (2003). "Consumption of fish and n-3 fatty acids and risk of incident Alzheimer disease." *Archives of Neurology* 60: 940-946.

Murthy, R. S., J. M. Bertolote, et al. (2001). *Mental Health: New Understanding, New Hope.* Geneva, World Health Organization.

Narrow, W. E., D. S. Rae, et al. (2002). "Revised prevalence estimates of mental disorders in the United States." *Archives of General Psychiatry* 59: 115-123.

National Institute on Alcohol Abuse and Alcoholism (2004). Alcoholism: Getting the Facts. *http://www.niaaa.nih.gov/publications/booklet.htm.*

National Institute on Drug Abuse (2004). Scientific Facts on Drug Abuse. *http://www.drugabuse.gov/drugpages/stats.html.*

Nemeroff, C. B. (1998). "Psychopharmacology of affective disorders in the 21st century." *Biological Psychiatry* **44**: 517-525.

Nemets, B., Z. Stahl, et al. (2002). "Addition of omega-3 fatty acids to maintenance medication treatment for recurrent unipolar depressive disorder." *American Journal of Psychiatry* **159**: 477-480.

Nesse, R. M. (1984). "An evolutionary perspective on psychiatry." *Comprehensive Psychiatry* **25**: 575-580.

Nesse, R. M. (1990). "Evolutionary explanations of emotions." *Human Nature* **1**: 261-289.

Nesse, R. M. (1991). "What good is feeling bad?" *The Sciences* **November/December**: 20-37.

Nesse, R. M. (1998). "Emotional disorders in evolutionary perspective." *British Journal of Medical Psychology* **71**: 397-415.

Nesse, R. M. (1999a). "The evolution of hope and despair." *Social Research* **66**: 429-469.

Nesse, R. M. (1999b). "Proximate and evolutionary studies of anxiety, stress and depression: Synergy at the interface." *Neuroscience and Biobehavioral Reviews* **23**: 895.

Nesse, R. M. (2000). "Is depression an adaptation?" *Archives of General Psychiatry* **57**: 14-20.

Nesse, R. M. (2005). Is the Market on Prozac? *http://www.edge.org/3rd_culture/story/100.html.*

Nesse, R. M., and A. T. Lloyd (1992). The evolution of psychodynamic mechanisms. *The Adapted Mind.* J. H. Berkow, L. Cosmides, and J. Tooby, eds. Oxford, Oxford University Press: 601-624.

Nesse, R. M., and G. Williams (1996). *Why We Get Sick: The New Science of Darwinian Medicine.* New York, Vintage.

Nestler, E. J. (1998). "Antidepressant treatments in the 21st century." *Biological Psychiatry* **44**: 526-533.

Noaghiul, S., and J. R. Hibbeln (2003). "Cross-national comparisons of seafood consumption and rates of bipolar disorders." *American Journal of Psychiatry* **160**: 2222-2227.

Osborne, L. (2001). "Regional disturbances." *The New York Times Magazine* **May 26**: 98-102.

Panter-Brick, C., and M. Smith (2000). *Abandoned Children.* Cambridge, Cambridge University Press.

Payer, L. (1988). *Medicine and Culture*. New York, Henry Holt and Company.

Peet, M., J. Brind, et al. (2001). "Two double-blind placebo-controlled pilot studies of eicosapentaeonic acid in the treatment of schizophrenia." *Schizophrenia Research* **49**: 243-251.

Pierpont, C. R. (2004). "The measure of America." *The New Yorker* **March 8**: 48-63.

Pinker, S. (1997). *How the Mind Works*. New York, Norton.

Porter, R. (1996). Mental illness. *Cambridge Illustrated History of Medicine*. R. Porter, ed. Cambridge, Cambridge University Press: 278-303.

Porter, R. (2002). *Madness: A Brief History*. Oxford, Oxford University Press.

Portin, P., and Y. O. Alanen (1997). "A critical review of genetic studies of schizophrenia II. Molecular genetic studies." *Acta Psychiatrica Scandinavica* **95**: 73-80.

Price, J. (1998). "The adaptive function of mood change." *British Journal of Medical Psychology* **71**: 465-477.

Price, J., L. Sloman, et al. (1994). "The social competition hypothesis of depression." *British Journal of Psychiatry* **164**: 309-315.

Price, J. S. (1967). "The dominance hierarchy and the evolution of mental illness." *Lancet* **2**: 243-246.

Puri, B. K., R. Steiner, et al. (1998). "Sustained remission of positive and negative symptoms of schizophrenia following treatment with eicosapentaenopic acid." *Archives of General Psychiatry* **55**: 188-189.

Reis, L. C., and J. R. Hibbeln (submitted). "Fish, gods and the psychotropic properties of omega-3 fatty acids." *American Journal of Clinical Nutrition*.

Richards, M. P., P. B. Pettitt, et al. (2001). "Stable isotope evidence for increasing dietary breadth in the European Mid-Upper Paleolithic." *Proceedings of the National Academy of Sciences* **98**: 6528-6532.

Ridley, M. (2000). *Genome: The Autobiography of a Species in 23 Chapters*. New York, HarperCollins.

Ridley, M. (2003). *Nature via Nurture*. New York, HarperCollins.

Rittenbaugh, C. (1982). "Obesity as a culture-bound syndrome." *Culture, Medicine and Psychiatry* **6**: 347-361.

Salem, N. (1999). "Introduction of polyunsaturated fatty acids." *Backgrounder* **3**: 1-8.

SAMSHA (Susbstance Abuse and Mental Health Services Administration) (2003). SAMSHA's National Mental Health Information Center. *http://www.mentalhealth.org/databases_exe.asp?D1=MA&Type=PC&Myassing=list.*

schizophrenia.com. *http://www.schizophrenia.com/index.html.*

Service, E. R. (1963). *Profiles in Ethnology.* New York, Harper and Row.

Shorter, E. (1997). *A History of Psychiatry: From the Era of the Asylum to the Age of Prozac.* New York, John Wiley and Sons.

Shorter, E. (2003). *Madness: A Brief History.* Oxford, Oxford University Press.

Siegel, J. T. (1986). *Solo in the New Order.* Princeton, N.J., Princeton University Press.

Silverman, J. (1967). "Shamans and acute schizophrenia." *American Anthropologist* **69**: 21-31.

Simons, R. C. (1996). *Boo! Culture, Experience, and the Startle Reflex.* Oxford, Oxford University Press.

Small, M. F. (1993). *Female Choices: Sexual Behavior of Female Primates.* Ithaca, N.Y., Cornell University Press.

Small, M. F. (1998). *Our Babies, Ourselves: How Biology and Culture Shape the Way We Parent.* New York, Anchor Books.

Small, M. F. (2000). "Aping culture." *Discover* **21**: 53-57.

Smuts, B. B., D. L. Cheney, et al. (1987). *Primate Societies.* Chicago, University of Chicago Press.

Spiro, M. E. (1967). *Burmese Supernaturalism: A Study in the Explanations and Reduction of Suffering.* Upper Saddle River, N.J., Prentice Hall.

Spiro, M. E. (1970). *Buddhism and Society: A Great Tradition and Its Burmese Vicissitudes.* New York, Harper and Row.

Stanford, C. (2003). *Upright: The Evolutionary Key to Becoming Human.* Boston, Houghton-Mifflin.

Stevens, A., and J. Price (1996). *Evolutionary Psychiatry: A New Beginning.* New York, Routledge.

Stoll, A. L., E. Severus, et al. (1999). "Omega 3 fatty acids in bipolar disorder." *Archives of General Psychiatry* **56**: 407-411.

Suomi, S. J., H. F. Harlow, et al. (1974). "Reversal of social deficits produced by isolation rearing in monkeys." *Journal of Human Evolution* **3**: 527-534.

Suomi, S. J., C. D. Eisele, et al. (1975). "Depressive behavior in adult monkeys following separation from family environment." *Journal of Abnormal Psychology* **84**: 576-578.

Suwanlert, S. (1976). *Phii pob*: Spirit possession in rural Thailand. *Culture-Bound Syndromes, Ethnopsychiatry, and Alternate Therapies*. W. P. Lebra, ed. Honolulu, University Press of Hawaii: 68-87.

Tanskanen, A., and J. R. Hibbeln (2001). "Fish consumption, depression, and suicidality in a general population." *Archives of General Psychiatry* **58**: 511-512.

Tattersall, I. (1993). *The Human Odyssey: Four Million Years of Human Evolution*. Upper Saddle River, N.J., Prentice Hall.

Teoh, J. (1972). "The changing psychopathology of amok." *Psychiatry Research* **35**: 345-351.

Tseng, W. (2001). *Handbook of Cultural Psychiatry*. New York, Academic Press.

Tseng, W.-S., M. Kan-Ming, et al. (1988). "A sociocultural study of *koro* epidemics in Guangdon, China." *American Journal of Psychiatry* **145**: 1538-1543.

Tsukada, H., T. Kakiuchi, et al. (2000). "Docosahexaenoic acid (DHA) improves the age-related impairment of coupling mechanisms between neuronal activation and functional cerebral blood flow response: A PET study in conscious monkeys." *Brain Research* **862**: 180-186.

U.S. Department of Health and Human Services, SAMSHA, Center for Mental Health Services, National Institutes of Health, National Institute of Mental Health (1999). *Mental Health: A Report of the Surgeon General.* Rockville, Md., National Institutes of Health.

van Schaik, C. P., M. Ancrenaz, et al. (2003). "Orangutan cultures and the evolution of material culture." *Science* **299**: 102-105.

Vega, W. A., B. Kolody, et al. (1998). "Lifetime prevalence of DSM-III-R psychiatric disorders among urban and rural Mexican Americans in California." *Archives of General Psychiatry* **55**: 771-778.

Walter, R. V., R. T. Buffler, et al. (2000). "Early human occupation of the Red Sea coast of Eritrea during the last interglacial." *Nature* **405**: 65-69.

Weatherall, M. (1996). Drug treatment and the rise of pharmacology. *Cambridge Illustrated History of Medicine*. R. Porter, ed. Cambridge, Cambridge University Press: 246-277.

Westermeyer, J. (1972). "A comparison of *amok* and other homicides in Laos." *American Journal of Psychiatry* **129**: 703-709.

Whitaker, R. (2001). *Mad in America: Bad Science, Bad Medicine, and the Enduring Mistreatment of the Mentally Ill.* New York, Perseus Books.

Whiten, A., J. Goodall, et al. (1999). "Cultures in chimpanzees." *Nature* **399**: 682-685.

WHO (World Health Organization) (2004). Mental Health: The Bare Facts. *http://www.who.int/mental_health/en/.*

Wilson, D. R. (1993). "Evolutionary epidemiology: Darwinian theory in the service of medicine and psychiatry." *Acta Biotheoretica* **41**: 205-218.

Wilson, D. R. (1998). "Evolutionary epidemiology and manic depression." *British Journal of Medical Psychology* **71**: 375-395.

Wright, O. (2003). MD's Hand Out 8 Million More Prescriptions for Depression, Anxiety and Stress. *www.timesonline.co.uk/newspaper.*

Yap, P. M. (1962). "Words and things in comparative psychiatry, with special reference to the exotic psychoses." *Acta Psychiatrica of Scandinavica* **38**: 163-169.

Yoshida, S., M. Miyazaki, et al. (2001). "Change of oligosaccharides of rat brain microsomes depending on dietary fatty acids and learning tasks." *Journal of Neuroscience Research* **63**: 185-195.

ACKNOWLEDGMENTS

IT'S A SIGN OF PROGRESS IN THE TREATMENT OF MENTAL ILLNESS IN OUR culture that so many people were willing to share their experiences, beliefs, and feelings about mental health and illness with me during the research and writing of this book. The fact that many people were also unwilling to let me use their real names is also a sign of how far we have to go.

I thought writing a book on mental illness would be depressing, but it wasn't. Everyone I interviewed was positive, hopeful, and caring. I am especially grateful to Bill Wilson for the most moving interview I've ever experienced, Joe Hibbeln for his infectious enthusiasm, and Bob De Luca for reciting Shakespeare and leaving me speechless.

Michelle Tessler is the agent every writer should have. Michelle believed in this project from the beginning and she stuck with the book, and with me, often navigating through stormy waters. Her advice on this book and other projects has been invaluable, and very much appreciated.

I am also grateful to my writer friends, the great wordsmiths Paul

Cody, Mike May, and Steve Mirsky. I don't think I would write a word if these guys weren't there to keep the writing life in perspective for me.

And thanks to my dear friend Ann Jereb, who is always wise about matters psychological.

My husband, Tim Merrick, deserves more than a book dedication. He deserves a medal.

INDEX

A

Aggressive/violent behavior
  access to weapons, 115
  amok behavior, 113–116
  cultural generalizations, 90–91
  diet and, 82, 85
  genetic and experiential influences, 67
Agoraphobia, 124, 126
Alcohol consumption
  bipolar disorder and, 13
  genetic risk in, 32–33
  prevalence of abuse, 14
  psychological effects, 15
Allan, Steven, 51
American Psychiatric Association, 1–2. *See also Diagnostic and Statistical Manual of Mental Disorders*
Amok behavior, 113–116, 128
Anorexia nervosa, 124–126

Anthropology, 149–151
Antidepressant drugs, 16, 153, 154–155
Antipsychotic medications, 16, 155
Anxiety
  cross-cultural manifestations, 127–128
  cultural change and, 153
  evolutionary model, 35–36, 46–50, 71–72
  genetic risk, 66–72
  koro, 119–122
  prevalence, 46, 66
  serotonergic system and, 68–71
  subtypes, 16
Arboleda-Florez, Julio, 114
Arctic cultures, 122–123, 128, 141–142
Ataque de nervios, 123
Attachment theory, 58–59
Attention deficit hyperactivity disorder, 126–127
Australian aborigines, 146
Azande people, 137–140

B

Balinese people, 100–101, 132–133
Barí people, 90–91
Beckerman, Stephen, 90–91
Behaviorism, 21
Benedict, Ruth, 105–108
Beng people, 99–100
Bennett, Alison, 68–69
Biology
    amok behavior and, 116
    culture and, 94, 102–103
    environmental interaction,
        93–94
    evolution of brain, 36–37
    evolutionary psychiatry model of
        mental health, 55
    fight-or-flight response, 47
    medical model of mental illness, 17,
        18, 152–153
    primate studies, 3–4
    serotonergic system, 67–71
    startle reflex, 116–117
    See also Diet
Bipolar manic-depression
    alcohol consumption and, 13
    clinical conceptualization, 12
    diet and, 82–83
    experiences of individual with, 8–
        13, 14
    standards of care, 12
Blame, 136, 137, 152
Boas, Franz, 93, 105
Bock, John, 109–110
Bowlby, John, 58–59
Brain fog, 123
Brief therapy, 23–24
Brooks, Alison, 86
Brumberg, Joan Jacobs, 124–126
Bulimia, 124, 125
Bushmen of Botswana, 143–145
Buydens-Branchy, Laure, 82

C

Caffeine, 88
Cannon, W. B., 47
Childbirth, 97–99
China, 121, 148–149
Chlorpromazine, 16
Client–practitioner relationship
    anthropological approach to
        psychotherapy, 149
    communication issues, 32
    cultural bias in, 31–32
Clinical trials, 154–155
Cognitive-behavioral therapy, 21
Communal societies, mental health
    care in, 30–31
Communication
    client–practitioner, 32
    culturally-mediated concept of
        mental illness, 103–104
    in transmission of culture, 103
Consciousness
    perception of reality, 42–43
    ritual alteration, 145–146
Cordain, Loren, 86
Cortisol, 82
Cost of care, 27–28
Coyle, Joseph, 45–46
Crawford, Michael, 85–86
Cultural characteristics
    change over time, 100–101
    cross-cultural comparison, 99–100
    definition of culture, 92, 94–95
    evolutionary development, 96–97,
        130–131
    formation and transmission, 91–92,
        94, 99–100, 102
    generalizations, 90–91
    globalization trends, 156
    historical development of concept
        of culture, 92–94
    primate studies, 95–96
    social interaction, 99

spiritual and supernatural beliefs,
132–135
See also Culturally-mediated
conceptualizations; Culture-
bound syndromes
Cultural relativism, 93
Culturally-mediated
conceptualizations
anthropological approach to health
care, 149–151
behavioral norms, 106–108, 111
childbirth, 97–99
diagnostic systems, 12
formation and transmission, 102–
108
mental illness, 2–3, 4, 33, 103–104,
109–110
resistance to pharmacotherapy, 18–
19
significance of, 31, 33–34
See also Culture-bound syndromes
Culture-bound syndromes, 122–124
clinical significance, 127–129
conceptual basis, 4, 110–112
diagnostic classification, 127
koro, 119–122
latah behaviors, 116–119
manifestations in Western culture,
124–127
running amok, 113–116

D

Dancing, 142–145
Darwin, Charles, 38
De Luca, Robert, 25–28
de Waal, Frans, 63
Deinstitutionalization movement, 26
Depression
anthropological approach to
treatment, 150
clinical features, 50

cross-cultural manifestations, 127–
128
cultural change and, 157
as culture-bound syndrome, 127
diet and, 74–76, 81–82
evolutionary psychiatry's
perspective, 39, 41, 50–54
gender differences, 154
goal-seeking and, 52–53
heart disease and, 82
hope and, 53–54
neurasthenia and, 148–149
placebo effect in pharmacotherapy,
154–155
poverty and, 153
prevalence, 81
serotonergic system and, 67–71
situational, 19–20
social context of psychological
development, 57–62, 65–66
Development, individual
benefits of anxiety, 71
cultural influences, 106–108
diet and, 84–85, 94
serotonergic system and, 69,
79–80
social relationships in, 57–59, 61–
62, 69, 72–73, 97–98
transmission of culture, 102–103
trauma experience, 104
*Diagnostic and Statistical Manual of
Mental Disorders,* 12, 127
Diagnostic categorization
anxiety and anxiety disorders, 46
cross-cultural commonalities, 127–
128
cultural mediation of, 12, 112
culture-bound syndromes, 127
possession by spirits, 136–137
racial/ethnic bias, 31
sociocultural context
considerations, 149–150
Western approach, 12, 43, 151–152

Diet, 4
  eating disorders, 124–126
  mind–body linkage, 78–80
  See also Fat, dietary
Dissociative disorders, 136–137
Docosahexaenoic acid, 80, 83, 85
Dopaminergic system, 80
Drugs. *See* Pharmacotherapy;
  Substance use

E

Eating disorders. *See* Anorexia nervosa;
  Bulimia
Eicosapentaenoic acid, 83
Emotional functioning, 37–38, 40, 63–
  64
Escobar, Javier, 29–32, 33, 154, 156
Evans-Pritchard, E. E., 137–140
Evolutionary development
  emotional functioning, 37–38, 40,
    63–64
  human brain, 36–37
  human culture, 96–97, 130–131
  human diet, 85–88
  material culture, 130–131
  startle reflex, 116
Evolutionary psychiatry
  conceptual basis, 3, 36–41
  diet and, 85–88
  model of anxiety, 35–36, 46–50, 71–
    72
  model of depression, 39, 41, 50–54
  model of schizophrenia, 41, 44–46
  treatment approach, 39, 40, 41, 54–
    55

F

Fainting, 122
Fasting, 126

Fat, dietary
  biological function, 76–77
  brain composition and function
    and, 78–80
  evolution and consumption of, 85–
    88
  mood disorders and, 74–76, 80, 81–
    83
  psychotic disorders and, 83–84
  serotonergic system and, 79–80
  types of, 77
  Western diet, 75, 78
Feinstein, Howard, 17–20
Fight-or-flight response, 47
Flinn, Mark, 104
Forero, Juan, 157
Freud, Sigmund, 20–21, 38

G

//*Gangwasi*, 143–145
Gaw, Albert, 113
Gender differences in depression, 154
Genetics
  anxiety and, 66–67
  cultural risk factors, 32–33
  environmental context, 70–72
  rationale for primate studies, 57
  research, 31
  schizophrenia risk, 43–44
  serotonergic system, 67–71
Ghosts, 133, 134–135, 143, 145
Gilbert, Paul, 40, 51
Goodall, Jane, 57, 95
Gottlieb, Alma, 98

H

Hagen, Edward, 51
Harlow, Harry, 57–61, 62
Heart disease, 82

Hibbeln, Joseph, 74–76, 78–80, 81, 87, 88
Higely, Dee, 64
Hope, 157
  clinical significance, 5
  common features of mental illness treatment, 4
  depression and, 53–54
  as source of personal transformation, 9–10
Huaorani people, 91
Hughes, Charles, 112
Hunting and gathering societies, 143

I

Iceland, 81, 156–157
Imitation, 117–118
Immigrant experience, 93–94, 111–112
  bias in mental health care, 31–32
  mental illness risk, 32–33
  protective factors, 32
Infant separation studies, 57–59
Innes, Sheila, 80
Inpatient care
  deinstitutionalization movement, 26
  historical development in Western culture, 26
  origins, 15
Interpersonal functioning
  attachment theory, 58–59
  brain evolution and, 36–37
  causes of depression, 51–52
  cultural influence, 99
  evolutionary psychiatry model of mental illness, 41
  perception of reality, 42–43
  serotonergic system and, 69
  social anxiety, 48

social context of psychological development, 57–66, 67, 72–73
social hierarchies, 62–63
sociocultural context of mental health, 149–151
witchcraft beliefs and, 140

J

Japan, 81, 123
Javitt, Daniel, 45–46
Jilek, Wolfgang, 121–122
Jilek-Aall, Louise, 121–122
Jung, Carl, 38

K

Kagan, Jerome, 66
Kirsch, Irving, 155
Kleinman, Arthur, 147–151
Koro, 119–122
Kraepelin, Emil, 110
!Kung people, 143–145

L

Landy, David, 123
Latahs, 117–118
Lee, Richard, 143, 144, 145
Lewis, Aubrey, 38
Loomis, Tamara, 22–24
Low, Setha, 127

M

Macbeth, 28–29
Malaysia, 113, 115, 117–118
Marks, Isaac, 49
McGuire, Michael, 44–45
Mead, Margaret, 93, 105

Medical model of mental illness, 146
   accomplishments, 151
   biological bias in, 152–153
   diagnostic categorization in, 151–
      152
   pharmacotherapy and, 153–154
   possible harms in, 156
   status in Western culture, 2–3, 151
   Western practitioner's perspective,
      18
Mellor, Jan, 83–84
Mental health professionals
   in non-Western cultures, 134–135
   perceptions of drug therapy, 17–20
   perceptions of talk therapy, 22–24
   Western medical hierarchy, 17
Mental illness
   cross-cultural commonalities, 4,
      127–128
   culturally-mediated perception, 3,
      4, 31, 33–34, 103–104, 109–110
   as defense, 40, 46–47, 48–49
   diet and, 82–85
   evolutionary model, 37–38, 39–41,
      54–55
   genetic factors, 66–72
   poverty and, 27–28
   prevalence, 11
   primate social development and,
      57–62
   public perception and
      understanding, 2
   rationale for cross-cultural studies,
      29–33
   risk for immigrants, 32–33
   supernatural causes, 134–135
   Western conceptualization, 2–3, 12,
      25, 43, 112, 146, 151–152
   See also Culture-bound syndromes;
      Medical model of mental illness;
      Treatment of mental illness;
      specific disorder
Myanmar, 133–135

N

National Institute of Mental Health,
   29–30, 31
Nats, 134–135
Nemets, Boris, 83
Nerves, 15, 127–128
Nervous breakdown, 128
Nesse, Randolph, 35, 40, 41, 52–54, 153
Neurasthenia, 148
New Zealand, 81
Norepinephrine, 66
Novak, Melinda, 61

O

Obesity, 126–127
Omega fatty acids, 75, 76, 77–78, 80,
   81–82, 83–86, 87–88
Opiates, 15–16
Owens, Sylvia de la Pres, 80

P

Panic attacks, 47, 127
Peet, Malcolm, 83
Pharmacotherapy
   bipolar manic-depression
      treatment, 12–13
   cross-cultural comparison, 30–31
   cultural resistance to, 18–19
   current culture of American
      psychiatry, 1–2, 16–17, 153
   efficacy, 154–155
   evolutionary psychiatry's
      perspective, 39
   long-term effects, 19
   origins and development, 15–16
   possible harms in, 155
   possible overuse, 153–154
   prescribing authority, 17

talk therapy and, 20
trends, 153
Western practitioner's perspective, 17–20
See also specific drug; specific drug type
Phenothiazine, 16
Phi pob, 135
Pibloktog, 123, 128
Pitt Rivers Museum, 130–132
Placebo effect, 154–155
Political context
amok behavior, 113–114
latah behaviors, 118
Possession by spirits
as cause of mental illness, 134–135
common features, 135–136
treatment, 136
Western conceptualization, 136–137
witchcraft and sorcery in, 137–140
Poverty, mental illness and, 27–28, 153
Power relations, 28, 118
spirit possessions and, 135–136
Prescribing authority, 17
Prevalence of mental illness, 11
alcohol and substance abuse, 14
anxiety, 46, 66
depression, 81
psychotherapy utilization patterns and, 25
schizophrenia, 43
trends, 11
Price, John, 47, 51
Primate behavior and biology
anxiety studies, 66–67
emotional functioning, 63–64
gene/environment interactions, 71–72
manifestations of culture, 95–96
rationale for medical and behavioral studies, 56–57
serotonergic system, 68–71

significance of, 3–4
social context of psychological development, 57–64
Prozac, 12
Psychoanalytic theory/therapy, 20–21
Puerto Rico, 123

R

Racial/ethnic bias in mental health care, 31
Reality, perception of, 42–43
Reis, Laura, 88

S

Sapirstein, Guy, 155
Schizophrenia, 16, 30, 102–103, 154
clinical conceptualization, 43
dict and, 83–84
evolutionary psychiatry model, 41, 44–46
as information processing disorder, 44–45
prevalence, 43
risk factors, 43–44
shamanism and, 142
Selective serotonin reuptake inhibitors, 16
Serotonergic system, 67–71, 79–80
Shamanism, 111, 134–135, 141–142
Shenjing shuairuo, 148–149
Simons, Ronald, 112, 116, 117, 119
Skinner, B. F., 21
Social anxiety, 48
Somaticization, 149
Soy oil, 87–88
Spiritual and supernatural beliefs
altered consciousness and, 145–146
causes of illness, 134
cross-cultural commonalities, 145

human propensity for, 132–134
mental illness as possession, 134–137
shamanism, 141–142
trance dancing, 143–145
Spiro, Melford, 134, 136
Startle reflex, 116–119
Stevens, Anthony, 47
Stoll, Andrew, 82–83
Substance use
diet and, 82
See also Alcohol consumption
Suomi, Stephen, 59–60, 65–66, 70–71
Suoyang, 120
Suwanlert, Sangun, 135

T

Taijin kyofusho, 123
Talk therapy
brief therapy, 23–24
conceptual basis, 20–21
efficacy, 21
in evolutionary psychiatry, 41
historical and technical evolution, 20–21
pharmacotherapy and, 20
practitioner's perspective, 22–24
status in Western culture, 14, 17, 22, 25
utilization patterns, 25
Thailand, 135
Thorazine, 16
Tompkins County Mental Health Building, 25–28
Trance states, 136, 141, 143–145
Trauma exposure, 104
Treatment of mental illness, 4, 157
anthropological approach, 149–151
common features, 4
in communal societies, 30–31
cultural factors, 33

evolutionary psychiatry's perspective, 39, 40, 41
exorcism, 136
health care system, Western, 25–28
historical development in Western culture, 14–17, 26–27
informant experiences of American system, 8–13, 14
international comparison, 30
shamanism, 141–142
socioeconomic context, 27–28
state spending, 27
Western conceptualization, 12
See also Pharmacotherapy; Talk therapy
Troisi, Antonio, 44–45
Tseng, Wen-Shing, 120
Tungus people, 141–142

V

Vega, William, 32–33
Vitamin A overdose, 123

W

Wallace, Alfred Russel, 113–114
Watson, J. B., 21
Weix, Gigi, 100–101
Western culture
childbirth experience, 98–99
concept of anxiety, 46
concept of mental illness, 2–3, 12, 25, 43, 112, 146
concept of spirit possession, 136–137
as cultural amalgam, 29
culture-bound syndromes, 124–127
diagnostic approach, 12, 43, 151–152
diet, 75, 78, 81–82

globalization trends and, 156
historical development of
    treatment approaches, 14–17
informant experiences of
    psychotherapy in, 8–13, 14
inpatient care, 26
medical model of health and
    disease, 151–154
pharmacotherapy, 1–2
power relations, 28
resistance to
    psychopharmacotherapy, 18–19
status of talk therapy, 14, 17, 22, 25

Wilson, Bill, 8–13, 14
Witches and sorcerers, 134–135, 137–
    140

Y

Yap, Pow-Ming, 111

Z

Zuni people, 105, 106